The Elements
of Power

The Elements of Power

*Gadgets, Guns, and the Struggle
for a Sustainable Future in the
Rare Metal Age*

DAVID S. ABRAHAM

Yale UNIVERSITY PRESS

New Haven and London

Published with assistance from the foundation established in memory of Philip
Hamilton McMillan of the Class of 1894, Yale College.

Yale University Press books may be purchased in quantity for educational, business,
or promotional use. For information, please e-mail sales.press@yale.edu
(U.S. office) or sales@yaleup.co.uk (U.K. office).

Set in Minion type by Westchester Publishing Services.
Printed in the United States of America.

Library of Congress Control Number: 2015942454
ISBN 978-0-300-19679-5 (cloth : alk. paper)

A catalogue record for this book is available from the British Library.

This paper meets the requirements of ANSI/NISO Z39.48-1992
(Permanence of Paper).

10 9 8 7 6 5 4 3 2 1

To my father, who never read even page 1 but who was with me along the way, and to my mother, who read every page again and again, and then again.

Contents

Preface ix

1. Metals, Metals Everywhere 1
2. National Struggles: Mineral Veins and Battle Lines 18
3. Corporate Hurdles: Monopolies and Investment Incentives 38
4. Production Difficulties: Acid Washes and Talent Drains 67
5. Trading Networks: Smugglers and Supply Hiccups 89
6. Tech Needs: The Electronification of Everything 115
7. Environmental Needs: Rare Metals Are Green 134
8. War Effort: Hard and Smart Metals 155
9. Sustainable Use: The Environmental Calculus of the Rare Metal Age 173
10. The War over the Periodic Table 194
11. How to Prosper in the Rare Metal Age 214

Notes 231
Acknowledgments 299
Index 303

Preface

We have quietly entered a new era, the Rare Metal Age. The products we use every day, from smartphones to cars, require a great number of hard-to-come-by metals, combined in increasingly complicated amalgamations. This book shows where the ingredients that underpin our society come from, how they get to us, and how they impact the environment. I wrote this book in the hope that you, the reader, will consider the scope of our dependence on these metals and will recognize that our technology, as well as our economic and climate security, comes at a cost. How we plan to pay will affect our future in ways many of us have not previously understood.

To develop this understanding, I've tracked the trail of these rare metals from mine to gadget and from gadget to recycled afterlife. I chatted with Japanese salarymen in smoke-filled, back-alley restaurants in Tokyo; feasted on lamb and beer with Chinese officials in Mongolian-style yurts; waded in the muck with miners off the coast of Sumatra, Indonesia; and visited factories in a formerly secret Soviet town that once refined uranium for nuclear weapons. Yet, although I traveled to some of the most rugged spots on the planet, my formal research started at a far more genteel place: Japan's Ministry of Economy, Trade and Industry in 2010.

At the time, I was a foreign researcher with the Council on Foreign Relations working from an internal cubicle on the ministry's eleventh floor. While my perch lacked a view, I had a ringside seat to one of the greatest Asian resource battles of the past generation. China cut off exports to Japan of a set of rare metals, called rare earth elements, during a territorial skirmish in the East China Sea. As I witnessed Japan's rapid capitulation to many of China's demands, I saw a new geopolitical trump card. The battle over resources, which started when the first person learned how to coax metal from stone, had expanded into a larger battle—a war over the periodic table.

My interest in rare metals was initially geopolitical. I spent years examining the nexus of natural resources and geopolitics at an energy trading company, a Wall Street firm, and the natural resource division of the White House Office of Management and Budget. I also managed a nonprofit focused on water in Africa. But I looked at commodities consumed in large quantities, and I learned that the small rare metal world is far more complex and removed from the scrutiny of commodities like oil, gas, and coal.

Through my research in a dozen countries and my meetings with hundreds of miners, traders, scientists, and policymakers, I soon realized that geopolitical tensions were only one of the impediments to meeting a country's resource needs. Our rare metal supply lines—which ensure that the right metal of the right grade reaches the right location at the right time and at the right price—have become a marvel of modern efficiency. But these supply lines are precarious, and our increasing global demands will soon stress them further. Our high-tech, green society is built on a wobbly foundation.

Many have written about the economic and social effects of our international supply lines. Articles on abuse and mistreatment of workers in sweatshops and factories now abound in the press. Business journals examine which countries profit from the globalized supply chain of our electronic gadgets. Even coverage of the world's electronic waste, burned in Africa and Asia, has now entered the global discussion.

This book builds on those reports by examining the supply chain of the hidden rare metals that make modern life possible. These metals permeate our lives, allowing buildings to soar and our televisions to show vibrant colors. And because these metals are critical in green technology as well, they are the seeds of our sustainable futures, but, as a society, we know very little about them. They are buried deeply in our products and are indistinguishable to us from more well-known metals like aluminum and iron. Every age has its resources: iron provided weapons; coal, oil, and natural gas give light and power. Now ingredients like rare earth elements, indium and tungsten are critical in ways many of us have failed to grasp. Indeed, these rare metals may well be one of the most underreported areas of research today, despite, as we will see, the increasingly economic and geopolitical advantage they confer.

It's not just the metals themselves that remain mired in obscurity and secrecy; companies that use these materials often conceal their use, behind the veil of patents and trade secrets. Even those working in senior positions in many high-tech firms are not aware of the materials on which their operations are built.

Although I found unsavory aspects of the industry—illegal trade, unscrupulous mining, and environmentally noxious metal processing—most people in the business are decent. They are striving to make a comfortable life for themselves and their families. But, compared to other industries with which I'm familiar, the rare metal world has an air of secrecy, leading to a low level of trust. While the mining and trading of these metals is not a lawless realm, one would do well in this arena to abide by the Russian axiom, "trust, but verify."

For this reason, one of the greatest challenges to writing this book was the lack of reliable statistics. Verifiable facts are hard to come by and often turn out not to be facts after all, but someone's guess that was repeated enough times to etch itself into the industry as truth. Information reported in popular media can be far from accurate. Statistics on market size vary wildly between industry sources. Even government trade figures fall short; for example, China's rare earth exports to Japan and Japan's rare earth import statistics rarely align. In this situation, the absence of statistics can be as telling as the statistics themselves. Interviews with knowledgeable sources and time spent at mining sites globally were invaluable to fill in gaps in my knowledge, but they inherently carry the risk of my seeing only a certain perspective.

In my quest to get to the inner circle of truth, I have done my best to be true to the sentiment I've heard in the market and have attempted to filter the information through an unbiased lens. Figures, though, may be more estimates than actuals. And verifying black market data and activities relies heavily on the participants themselves and so data can be biased. The assumptions I make are my own. I have tried to make things simple, for example, by standardizing terms when possible; I have used metric tonnes for the weight measurement, but employed the U.S. spelling of the word "tons." I also tried to use measurements standard for the context; metals are sold by the ounce, flask, pound, and kilogram. Statistics in the book are as current as what was available at the time of writing, but since the industry publishes statistics infrequently, they may be less current than desirable.

Throughout the book, I use the term "rare metal" to refer to a set of metals mined in small quantities, often less than thousands of tons annually. Their use is, well, rare. However, there are also those metals that are geographically rare (such as tellurium) and another group of metals called "rare earth metals," which, although not synonymous with the term "rare metals," are a subset of them. The challenge was to use a term that conveyed limited use of these metals in relation to the use of other metals but also to emphasize their importance.

I also use the term "minor metal," which is an industry standard description for metals in limited production and a recently coined phrase "critical material" synonymously with "rare metal." I take solace in the fact that the metal industry is not actually very good at labeling metals. The term "precious metal" refers to silver, but not to metals that can actually be more precious such as germanium or terbium. I also use "acids" as shorthand to refer to the complex chemicals used in metal processing, but this also includes emulsifiers, flocculants, and other agents.

The book remains a work in progress; my time with it ended in early 2015. There is much more to write about recycling, processing, and the scientific properties of rare metals. As an observer, writer, and someone who dabbled in economics, the focus of the book may be short on scientific details for some readers. To be sure, people spend their entire careers understanding the properties of just one element, the mineralogy behind one ore, and the refining process of a single metal powder.

My hope is that this book introduces you not only to the hidden ingredients in our high-tech, green, and military lives but also to the characters and stories behind them, and that it explains the role of rare metals, including rare earth elements, in our products and describes how future demand for these re-sources can shape the global economy and geopolitics. This book comes at a defining time when rare metals are increasingly critical for high-tech, green, and military applications. Yet despite their prevalence, they are not understood. Just as cars made oil a staple of modern society, many of today's unheralded metals are likewise transformative to the products they find themselves in. This means that the rare metal market is one that demands far more scrutiny.

As you read this book I also hope you will reflect on these elements' power and promise, and the extent to which they are already embedded in your life. I hope you will see that the ability to harness the power of rare metals to make smartphones, for example, is as impressive as the phones themselves. It's not hyperbole to state that the fate of the planet and our ability to live a sustainable future in which technology can freely flow to the billions who do not yet have access depends on our understanding and production of rare metals and our avoidance of conflict over them.

I

Metals, Metals Everywhere

Microsoft CEO Steve Ballmer was incredulous. "There's no chance that the iPhone is going to get any significant market share. No chance," Ballmer prophesied during a CEO Forum before Steve Jobs released the iPhone in June 2007. But, by the end of the first week of sales, most storeroom shelves were bare; Apple and its AT&T partner sold hundreds of thousands of phones. The company was fast on its way to taking more than 20 percent of the smartphone market within just a few months.[1]

To those who waited in line outside Apple stores for a day or two to snap up the first phones—or paid others hundreds of dollars to wait for them—the iPhone was a revolution, the stuff of dreams. Although smartphones had been out for a few years, Jobs's phone, they believed, was set to be the smartest. Some in the media labeled it the "Jesus Phone" because of the religious fervor surrounding its launch and the blind faith that Jobs's new gadget would create not just a better phone, but a product that would reshape everything that followed it.[2]

As is now legend, the phone banished number buttons and the physical keyboard. Instead the iPhone became the first mainstream product to rely on a "multi-touch" glass screen—a highly functional screen that allows the tapping, sliding, and pinching that are now second nature for writing e-mails, determining directions, and hailing a cab. Jobs himself commented, "It works like magic."[3]

While Jobs's creative genius is beyond mythical, something greater was at play. Lost in the hubbub of new features and beyond the phone's powerful yet simple design was the most remarkable characteristic of iPhone—the reason why a powerful device can sit comfortably in the palm of your hand— it relies on nearly half the elements on the planet.[4]

These metals are the reason that devices are getting smaller and more powerful. For Jobs, the "magic" in his glass is due to a dash of the rare metal indium, which serves as the invisible link, a transparent conductor between the phone and your finger. A dusting of europium and terbium provides brilliant red and green hues on the screen, specks of tantalum regulate power within the phone, and lithium stores the power that makes the phone mobile. Rare metals are also crucial to manufacturing the iPhone's components: cerium buffs the glass smooth to the molecular level.

To be sure, the iPhone was far from the first or only product to rely on rare metals. In fact, the increasing use of rare metals correlates well with the sale of Apple and other computer products, which began some thirty years earlier. But Jobs's drive for smaller, more powerful gadgets led his company to increasingly harvest the complete palate of materials on the periodic table and deliver them to the masses. What's more, the iPhone's commercial success transformed our expectations of our gadgets. It spurred new industries, including mobile apps and tablets, making the power of rare metals indispensable not just in smartphones but in a myriad of new technologies. Jobs not only lived up to his word of reinventing the phone, he helped reinvent the world's resource supply lines. In the process, he also helped bring forth the dawn of a new era: the Rare Metal Age.

Rare metals are everywhere—really everywhere—from soaring bridges to earphone buds. They are in couches, camera lenses, computers, and cars. But they are rarely used alone or as the primary material. In essence they fill a role similar to that of yeast in pizza. While they are only used in small amounts, they're essential. Without yeast there's no pizza, and without rare metals there's no high-tech world.

We lack awareness of them because we never directly buy them as we do other commodities such as gas or corn. Rare metals are buried away in components that are essential to almost every gadget we use, like the rare earth permanent magnet. While the production of permanent magnets is approximately a mere $15 billion market today, if we were to add together the value of all industries that rely on these magnets— automobile, medical, and military—the sum would reach trillions of dollars.[5]

To paraphrase a slogan of the chemical corporation BASF, rare metals don't make the products we buy; they make the products we buy smaller, faster, and more powerful.[6] They made Jobs's iPhone thinner, more functional, and more mobile. This is because each rare metal has its own characteristics that serve very specific functions. For example, it can be malleable (indium), ductile (niobium), toxic (cadmium), radioactive (thorium), or magnetic (cobalt), or it can melt in your hand (gallium). And like characters in the X-Men comics, they all have their own superpowers. Terbium produces more vibrant light in television; dysprosium and neodymium make incredibly strong magnets possible; antimony helps resist fire.

Among the elements in the periodic table, roughly two-thirds are metals or metalloids, elements like silicon that share some characteristics of metals and nonmetals, and are most valuable because of their semiconducting properties. Of these, mines produce millions of tons each year of the best-known metals, like copper and zinc, which are called "base metals." Others, like gold and silver, have retained value for centuries, hence their name "precious metals."

Rare metals are in an umbrella category for almost all other metals. Their defining feature is that they're consumed in small quantities, hence "rare" when compared to base metals. On average, the world consumes individual rare metals in the hundreds or thousands of tons annually—the annual production of each can fit into just a few rail cars. By comparison, miners produce about 1.4 million tons of copper annually. According to data from the U.S. Geological Survey, if you add up the annual consumption of all materials that are considered rare, the amount would be substantially less than the quantity of copper consumed every year. The label "rare" does not mean that these metals are all geologically scarce. Indeed, some are plentiful. Others are abundant but seldom found in concentrations high enough to be mined profitably. To complicate labeling matters further, some in the industry call them "advanced" or "technology" metals because of their prevalence in electronic applications. Others call them "strategic" or "critical" because of their irreplaceability in their applications. Those who trade these materials alternatively call them "minor." Throughout the book I use these previous words interchangeably. (I also use the term "materials" instead of "metals" because in many cases it's not the pure metal that's traded, but a less-refined derivative of it.)[7]

Rare metals also encompasses rare earth elements, a set of seventeen atomically similar metals, which gained international attention in 2010, when fears of Chinese monopolistic control of production and export restrictions drove prices up nearly tenfold. Rare earth elements are a mere subset of rare metals but they share many of the same market dynamics. For example, many rare metals, like rare earths, must undergo challenging refining techniques. They are also traded in backroom deals rather than on open exchanges like other commodities such as oil.

If naming them is a challenge, classifying which metals are "rare" is even more problematic. Even the Minor Metals Trade Association, the organization that trades these metals, lacks a standard definition. By its count, members now trade forty-nine rare metals as in Figure 1, up from eight just three decades ago when manufacturers bought only a handful of them. (Many insiders can't even agree as to what is a rare metal and quibble about whether a specific metal should be labeled as such.)

But don't let the lack of an encompassing term or the small production levels fool you into underestimating their economic and geopolitical importance. These tiny quantities of metal have fostered incredible technological change. Rare metals are the base of our modern high-tech, green, and military industries. Rare metals are no less transformative than oil or coal. They will increasingly deserve the same attention

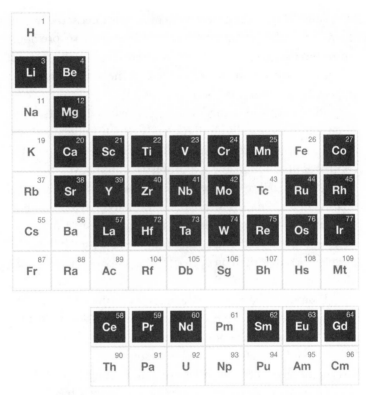

Figure 1. Minor Metals Trade Association's list of minor metals.
Courtesy of Minor Metals Trade Association (MMTA).

we afford fossil fuels, meaning those who control and manage their production and trade will increasingly reap outsized economic and geopolitical fortune. And yet, unlike oil or coal, they are often more limited in supply and far more complex to produce, and they originate from just a few places on earth. Many have such unique properties and uses that they cannot be switched out for cheaper or more abundant

									2He
			5 B	6 C	7 N	8 O	9 F	10 Ne	
			13 Al	14 Si	15 P	16 S	17 Cl	18 Ar	
28 Ni	29 Cu	30 Zn	31 Ga	32 Ge	33 As	34 Se	35 Br	36 Kr	
46 Pd	47 Ag	48 Cd	49 In	50 Sn	51 Sb	52 Te	53 I	54 Xe	
78 Pt	79 Au	80 Hg	81 Tl	82 Pb	83 Bi	84 Po	85 At	86 Rn	
110 Ds	111 Rg	112 Uub	113 Uut	114 Uuq	115 Uup	116 Uuh	117 Uus	118 Uuo	

65 Tb	66 Dy	67 Ho	68 Er	69 Tm	70 Yb	71 Lu
97 Bk	98 Cf	99 Es	100 Fm	101 Md	102 No	103 Lr

alternatives. Our reliance on rare metals is not just an abstract geopolitical issue or a topic of concern only to material scientists. It is a potential source of conflict. But it was not always this way.

Only some 150 years ago, nearly all of the materials in a person's home originated from a nearby forest or quarry. By the 1960s, with more developed supply lines and an increased demand for consumer appliances, the average American home used around twenty elements. Since then, material scientists have led a quiet revolution transforming the products we use and the materials that allow them to work. In the 1990s Intel used only fifteen elements to build its computer chips. Now the company demands close to sixty elements.[8]

The transformations in the products we use appear subtle to the untrained eye. Modern lights, for example, emanate a hue slightly different from that of their predecessors. But these slight changes in tone mask a profound change in resources. Whereas Edison's lightbulb contained a simple metal filament, the resources in today's LED lights are more akin to computer hardware, powered by gallium, indium, and rare earth elements.[9] This new set of elements and the applications that sprang from them, makes the products of today far more sleek than those of a generation ago. In the 1980s, when Maya Lin, the designer of the Vietnam Memorial in Washington, DC, asked Steve Jobs why he made clunky computers instead of thinner ones, he responded that he was waiting for the technology to build them.[10] What he really waited for was the time when material scientists would unlock the properties of rare metals and bring forth the flat screen.

Today, our daily individual purchasing decisions and the technology we use have significant ramifications on our rare metal supplies. Unfortunately, we have thought little about that connection between ourselves and these resources. While rare metals have been around since the beginnings of time, most were just discovered in the past few hundred years, and some just in the past century. Today, companies are using elements that scientists dismissed as mere impurities decades ago. Over the past thirty-five years, mining companies have produced four times more of many, if not all, rare metals than they produced from the dawn of civilization until 1980.[11]

It is the properties of rare metals like neodymium and dysprosium in the hardware of our gadgets that form the bedrock of new services that have revolutionized our lives. The media shower praise on the Silicon Valley innovators but the credit for our tech existence must be shared. What makes technologies from Google to Alibaba work is the proliferation of rare metal–laden technologies in our pockets. That so many people have smartphones generates new markets. But without decades of work by nameless mining engineers, metallurgists, and material scientists, Uber and Facebook would never have become household names. (Therefore, it's hardly ironic that, in the 1980s, Jobs bought the house of a mining and metallurgical engineer who had earned his wealth more than fifty years before him.[12])

These rare metal digital technologies have transformed not only the ways we travel, communicate, and shop but also our expectations. We have come to demand that technologies will become cheaper, more accessible, and more advanced each year—and that they do far more than many once thought possible. Although the multiple functions of our new gadgets appear to come with the opportunity to use fewer raw materials—after all, the iPhone is a computer, book, and music player—the reality is we use far more total resources. We don't realize this dynamic because we pay little attention to the increasing complexity inside our gadgets. Nor do we understand the tenuous supply lines that support our habits. Many of us simply wait for the next version of the iPhone and line up to buy it. Few understand the wondrous properties of rare metals that have made small, powerful devices inexpensive enough that billions of people can afford them. Rare metals now underpin our lifestyle. Indeed entire industries, like apps, and the foundation of many economies, are built on them.

Look at the influence of Jobs's phone: iPhone sales are now so large that analysts state that they increase the GDP of the United States and Taiwan, where many of the parts are made, as much as half of a percent. And they have resurrected Apple's fortunes. By 2012, sales of the iPhone totaled more than half the company's revenue. Add in its sibling, the iPad, and around three-quarters of the company's revenue springs from the iPhone technology. The invention transformed Apple from the eighty-fifth biggest company in the world, to the largest, the very largest, in just five years, overtaking ExxonMobil. Apple's ascendancy over Exxon reflects a new reality: the world is fast becoming as dependent on rare metals as it is on oil.[13]

Society was once bifurcated: the wealthiest 20 percent histori-cally consumed more than 80 percent of the resources. Or as the anthropologist Jared Diamond put it in 2008, each person living in a developed country consumed thirty-two times more material than one in a developing country.[14] When most of the world lived in developing countries, seemingly endless re-sources flowed to wealthiest ones. But that is changing—not because the wealthy are consuming less, but because everyone is consuming more. This is in part because wealthy countries have adapted to a high-tech, yet disposable lifestyle. They have paved the way for the entire world to live similarly without considering the global resource implications.

When research took me to Jakarta, Indonesia, I was struck by the number of cranes dangling over towering, hol-low concrete shells. These budding high-rises peered over slightly shorter ones constructed just a few years before. These new buildings will provide homes for increasingly wealthy, upwardly mobile Indonesians who are striving to live the technological-advanced dream, with all its attendant accoutrements. The people of Indonesia will be no different from the billions of others in the developing world, from South America to China, who are heading toward the same resource-intense existence. This means the global demand for metals, especially rare ones, will increase as countries follow a well-worn economic path. As people move from farms to better-paying jobs in cities, which necessitates the building of bridges, subways, and more power plants to supply electricity to charge people's smartphones and laptops, metal demand skyrockets.

How much more is needed? No one is sure. But the growth of steel consumption may be a relevant proxy. In South Korea, steel demand per person grew more than fivefold as the income of individual Koreans increased from $2,000 to $20,000. Likewise in China, the country consumed about one kilogram of steel per person when 30 percent of the population lived in cities, but jumped to more than five kilograms per person when 50 percent lived in urban areas, much of this for infrastructure. Because China plans for 70 percent of its people to live in cities by 2030, this means China will need even more steel. While this infrastructure needs more rare metals, the wealthier city can afford more gadgets which will compound demand for rare metals.[15]

Based on our current rate of rare metal resource production and our consumption patterns, we won't have the dysprosium necessary to build magnetic resonant imaging (MRI) machines; yttrium critical for military radar; or the tungsten for oil exploration drill bits. New high-tech inventions will only add urgency to expand our limited supply chains, meaning the future supply of materials for our gadgets is at stake. Numerous recent government and think tank studies highlight the risk of shortages over the next decade and some even longer. The American Chemical Society finds that over the next century, forty-four of the ninety-four naturally occurring elements face supply risks. While production levels of many elements will rise to meet demand, their report highlights a real concern.[16]

The future of our high-tech goods may lie not in the limitations of our minds, but in our ability to secure the ingredients to produce them. In previous eras, such as the Iron Age and the Bronze Age, the discovery of new elements brought forth seemingly unending numbers of new inventions. Now the combinations may truly be unending. We are now witnessing a fundamental shift in our resource demands. At no point in human history have we used *more* elements, in *more* combinations, and in increasingly refined amounts. Our ingenuity will soon outpace our material supplies.

This situation comes at a defining moment when the world is struggling to reduce its reliance on fossil fuels. Fortunately, rare metals are key ingredients in green technologies such as electric cars, wind turbines, and solar panels. They help to convert free natural resources like the sun and wind into the power that fuels our lives. But without increasing today's limited supplies, we have no chance of developing the alternative green technologies we need to slow climate change.

Our demands are now pushing against the bounds of what we can sustainably produce. Fluctuations in the complex supply line will affect society in unpredictable ways. New supplies of dysprosium could speed the development of highly efficient wind turbines, and conversely, a lack of it could drive up the cost of hybrid vehicles. It's no understatement to say that our use of rare metals will determine the fate of the planet.

Back in the Iron Age, a new set of strong iron weapons enabled those who mastered the art of turning stone into metal to conquer their neighbors. Today, although the weapons have changed, the situation is much the same: those who can master the elements make stronger foes. The difference is that whereas throughout history people have relied on metals to make weapons such as swords stronger and harder, today rare metals make weaponry smarter. Consider, for example, new missile weapon systems like Israel's Iron Dome, which gives a glimpse of what may be possible over time.

When a rocket flies toward an Israeli town, the system's computers and sensors decide within milliseconds when and where to launch a precision-guided missile to intercept it. By 2014, the Iron Dome had reportedly knocked out upward of 85 percent of the rockets headed toward Israeli cities, saving an untold number of lives and altering the nature of warfare.[17]

A few decades ago, a system like the Iron Dome, as well as any precision-guided missiles systems and even drones, were the stuff of science fiction. Today's technology relies on advances in radar, computers, and guidance systems. And at the core of each component lies rare metals. While the exact components of the Iron Dome are classified, the system, as any such sophisticated system today, must make abundant use of rare metals. Its computer screens use indium; rare earths are in the fin actuators that guide the missiles; and microchips full of rare metals drive their computers. These metals underpin complex weapons systems and ultimately a country's national defense.

For years, companies and countries took their rare metal supply lines for granted, unaware of the material makeup of their products. In fact, in 2011, Congress forced the U.S. military to research its supply chains because the Pentagon was having difficulty determining which advanced metals it needed.[18] As the materials that make up product components have become more varied and complex, those who rely on sophisticated hardware can no longer afford to remain in the dark.

Now, corporate and government leaders are realizing how important rare metals are. Indeed, efforts to secure rare metals have sparked a war over the periodic table. In offices from Tokyo to DC, in research and development labs from Cambridge, Massachusetts, to Baotou, China, and in strategic command centers the world over, new policies and the launching of research programs are ensuring that nations have access. The struggle for minor metals isn't imminent; it's already here and is shaping the relationship between countries as conflicts over other resources did in the past.

Just as the Cold War split the world along ideological lines, this new struggle will create fissures between those who have access to rare metal resources and those who do not. Because whole industries are built on just a few rare metals, disruptions to their supply can have profound global implications and give countries tremendous leverage. Billion-dollar companies are often beholden to just one country such as Congo or Kazakhstan—or even one particular mine—for a vital advanced metal. For most companies it's "difficult, if not impossible, to trace the minerals' origins," as the computer manufacturer Dell notes. Such lack of transparency is hardly a fail-safe situation when companies often need hundreds or thousands of components or more.[19]

This book is the first of its kind to explain what rare metals are, where they come from, and how they are used. We will meet miners, investors, and material scientists as we explore these metals and the seemingly unlimited opportunities they offer us. You will understand minor metals' complex supply web that begins on a hillside in Chile or on the edges of a Congo jungle, where people toil with simple shovels and picks, and ends on your desktop, in your pocket, or on a military base.

We will see the increasing importance of rare metals through the past century. Just as dependence on oil forced many countries into uneasy relationships with oil-rich regimes, now metal-rich countries like Russia and China have a new hold on trading partners. And some have begun to flex their muscles as China did in 2010, when it restricted the export of rare earth elements to Japan. In fact, the role of China and its control over rare metals is an important theme that plays throughout this book.

We will track the journey of these materials from rock to metal and come to understand how rare metal supply lines, which may appear similar to those of iron or oil, are far more complex. They are often dominated by a few entrenched suppliers and can take more than a decade to establish. This means that although higher prices will often bring on new supply, the principles of supply and demand do not work in a timely fashion. Inventions of new high-tech products can create resource demands far more quickly than suppliers can increase supplies, leading to price spikes.

What's more, because the cost of starting new projects is exorbitant, junior mining companies often spend more time looking for financial resources than for mineral ones. But those are just the first obstacles. Receiving regulatory approvals, creating the refining techniques to coax the metal from the minerals, and predicting demand can vex even the savviest mining company executive.

When companies can overcome those hurdles and produce rare metals, their material enters a channel of small trading shops where secrecy reigns and a reliable delivery is prized. A transparent market benefits those who produce the metals and those who buy them because it establishes clear prices, but it hurts the trading business. Traders' profit comes not only from the metals they peddle but also from their monopoly of information. There are profits in obscurity.

The future market looks bright for these traders, as we will see, because high-tech goods are getting cheaper, green products are more in demand, and countries are spending more on defense. The proliferation of cheaper rare metal–laden technology is colliding head-on with the increasing purchasing power of people in even the poorest communities. This can create surreal scenes—in places that have no clean water or paved roads, people have phones and TVs—but it also sows the seeds of a high-tech lifestyle in new lands.[20]

As demand for these rare metals grows, it is important to understand the environmental and geopolitical effects of increased production. Whereas the total environmental impact of producing rare metals is small in comparison to producing traditional commodities, the impact per kilogram (or pound) is far greater because of the quantity of chemicals and energy needed to refine the metals. And with little oversight of operations in some countries, the production of rare metals can be ruinous to the surrounding communities. Despite the environmental challenges, it's a risk that countries like China take because rare metal production will confer economic and geopolitical benefits that were previously reserved for more traditional commodities. As we will see, some countries, such as Japan and Germany, are realigning their relationships to ensure a reliable supply of rare metals.

The challenge is to produce and use rare metals efficiently while at the same time developing a sustainable supply chain. It is something we can do only if we both understand our use and dedicate ourselves to thinking about solutions, as this book does in its final pages. Otherwise we will repeat past mistakes: we once averted our eyes from the challenges of relying on fossil fuels, and now we are at risk of ignoring the dangers of relying too much on too little. And there are glimpses of the challenges that lie ahead, especially when the world relies on very rare metals from very remote places.

II
National Struggles
Mineral Veins and Battle Lines

During his thirty-one years as Zaire's president, Mobutu Sese Seko plundered the country's resources. He amassed billions of dollars to fund a lavish lifestyle and to dole out to supporters. In 1978, almost thirteen years after coming to power, when Soviet-backed rebels from Angola seized the Katanga region, known for its rich cobalt deposits, and challenged his control, Mobutu was quick to send in the military.

The ensuing battle choked off cobalt supplies to much of the world. The price of the rare metal—essential for permanent magnets in electric motors and heat-resistant alloys in jet engines—spiked to more than $60 per pound from $10 per pound in less than a year. Manufacturers scrambled for supplies. With the price so high, in fact, producers found it profitable to transport the silver-blue metal by air from the metal processors, when traditionally they had shipped it by boat.[1]

Rumors spread about greater geopolitical battles at play beyond the jungles of Africa. A few months before the rebels seized Katanga, the Soviet Union bought massive quantities of cobalt from Congo for its military-industrial needs. This move surprised many metal traders and simultaneously raised fears in government circles, especially in the United States, which relied on Zaire for 40 percent of its cobalt. Other rumors swelled that the Soviets wanted to take over the global cobalt market, hoarding supplies and crippling U.S. industries. A few years later, the secretary of state, Alexander Haig, said he thought the Soviet Union had started a resource war, and Zaire was the first battle.[2]

Some industries, such as paint manufacturing which used cobalt for pigment, shifted away from using it because substitutes were easy to find. Others tried to reengineer their products to eliminate cobalt, a strategy that often meant using less effective materials and led to higher prices for inferior products. The shortage spooked military planners and aviation executives since they had no substitute for cobalt alloys in new jet engines and other military applications.[3] But it may have been cobalt's use in permanent magnets that created some of the greatest challenges.

First used in the 1960s, cobalt permanent magnets quickly found their way into military applications such as microwave communication systems.[4] Over the next decade, they made their way into small motors because they were tiny and could be formed into a variety of shapes. Permanent magnets are useful to system designers because they sustain a strong fixed charge over an indefinite period of time while expending no energy in the process. At the time, no one knew whether cobalt magnets could be replaced, and if so, with what.

The fact that war disrupted mineral trade is neither surprising nor new. But it was new that a small insurgency, in a far-off land wreaked havoc upon the world's largest companies by cutting off most of their supply of a seemingly trivial metal.

The cobalt fight highlighted the tenuousness of vital military supply lines, which had come to rely on rare metals. Fortunately for those who relied on cobalt, Mobutu regained control of the mines, and simultaneously a global recession reduced demand for cobalt and reduced its price. But the conflict had started a scramble to replace cobalt.[5]

The widespread use of a rare metal whose production comes from an unstable source, as in the case of Zaire, may seem shortsighted, but what is myopic to the diplomat can be genius to the material scientist. The decisions about which ingredients to use in our devices come down to which materials can solve technical challenges, not which metals are in abundant supply.

This was the case when Masato Sagawa began work as a junior research scientist at the Japanese electronics giant Fujitsu, about five years before the Congo conflict. Although he had never taken much interest in magnets during his doctoral studies, let alone cobalt, his task at Fujitsu was to strengthen a samarium-cobalt magnet that kept chipping. Over the course of the project, the young scientist grew enamored with magnets, staying late at the lab to better understand their composition.

He knew cobalt alone couldn't create a permanent magnet, but mixing it with the rare earth element samarium could. Samarium helps form a unique crystalline atomic structure that assists in aligning smaller magnetic fields of individual atoms. This creates a strong, permanent magnet.

Sagawa thought that he could apply the same theory to make a permanent magnet out of iron and a more abundant rare earth, neodymium. This new magnet wouldn't face the price and supply limitations of cobalt and samarium. He believed it might even be stronger than existing magnets because iron had a more powerful magnetic disposition than cobalt.[6] Sagawa started working on his neodymium-iron magnet in 1976, spending weekends and nights at work and rarely seeing his newborn child. But by January 1978, he still had no success and the obstacles appeared formidable.

Masaaki Hamano, a leading permanent magnet researcher of the day, commented during a conference Sagawa attended that the iron–rare earth magnet, like the one Sagawa was working on, was impossible because the distance between the iron atoms was too small to allow for proper spacing to form a magnet.[7] But this presumed impossibility inspired Sagawa's solution—he would create more space between the iron atoms by adding boron to the rare earth–iron mix. With its smaller atomic size, boron increased the molecular distance between atoms by shoehorning itself in the spaces between them.[8]

As the war raged in the Congo, the newly inspired Sagawa continued his efforts while unbeknownst to him, materials scientists in corporate and military labs were blazing a similar trail to reduce the world's reliance on Zaire's cobalt.[9] But Sagawa's concern at the time was more parochial: not competition, but temperature.

Magnets are finicky—if they get too warm, beyond what is known as the Curie temperature, they lose their magnetism. Sagawa's magnets were losing their magnetic properties at such a low temperature that it would preclude their use in (hot) motors. Sagawa quickly discovered that removing a little neodymium and adding dysprosium helped his magnet to retain its magnetism over a wider range of temperatures—up to 310 degrees Celsius. This was high enough to allow for a broad range of potential applications.[10] He'd done it: created a strong, permanent magnet without using cobalt or samarium. There was just one problem.

Dysprosium was a rare earth element, geologically scarcer than even samarium or cobalt, and with far more significant supply-chain problems of its own. Few places produced the material. Sagawa told me its addition to the mixture was meant to be temporary. Relying on such a rare metal produced in large part in China, he thought, would be problematic.[11] He was right.

A little past 9:30 on the morning of September 7, 2010, a Japanese Coast Guard vessel in the East China Sea spots a large Chinese fishing trawler off the coast of islands, known as Senkaku in Japanese and Diaoyu in Chinese. This was not the first time Chinese boats had entered an area governed by Japan. In fact, this was slowly becoming a common occurrence that stretched back to 1978, when thirty-eight Chinese fishing boats, some outfitted with automatic weapons, sailed into the disputed area.[12]

The Japanese have little tolerance for incursions in the Senkakus as their claim to the islands goes back to 1895 when the country annexed them following the Sino-Japanese War. When you ask Japanese officials about the territorial dispute, they will look at you askew; for them, there is no territorial dispute. "It's our land," one government official told me. It was almost insulting for him to answer my question, similar to asking an American diplomat whether Hawaii is part of the United States.

But more recently China has asserted claims to these islands extending hundreds of years earlier.[13] Both countries support their assertions with different maps and nationalist zeal that allow little wiggle room for a diplomatic solution. The dispute is wrapped up in a morass of misunderstanding, one-upmanship, and an eye toward resource security due to the rich resources nearby.

On that morning, the Japanese vessel named the Yonakuni pulls alongside the thirty-seven-meter Chinese trawler blaring messages to the crew in Chinese from loudspeakers, "You are inside Japanese territorial waters. Leave these waters." As videos later leaked by a naval officer that day show, instead of leaving, the Chinese boat bends toward the stern of the larger Japanese cutter, hitting it and sailing on. Forty minutes later, another Japanese boat pulls beside the Chinese trawler to find a handful of people on deck and the forty-one-year old shirtless captain standing outside the bridge with a cigarette dangling from his lips. The drone of sirens, the demand for the boat to leave the area, and a rolling video camera onboard the Japanese vessel had no effect. The captain enters the bridge and once again veered the trawler into the Japanese ship, causing a larger collision.[14]

Since World War II, the Japanese government has attempted to balance diplomatic sensitivities with firm defense of its territory. Tokyo has been able to manage these incursions with little fanfare. In previous cases, Japan either escorted the ships away from the island or took the offenders into custody and quickly deported them.[15] But, in 2010, the Japanese response was different. The Democratic Party of Japan had recently come to power and was not well versed in international protocols. While the previous ruling party, the Liberal Democratic Party had, over the years, ruffled the feathers of Chinese leadership with provocative actions, it had avoided high-stakes territorial brinkmanship during its fifty-five years of virtually uninterrupted rule.[16]

The new leadership initially detained the trawler's crew and captain. Authorities released the crew after a few days, but planned to put the captain on trial. Chinese leadership demanded the immediate release of the captain and, in retaliation, detained four Japanese citizens in China, and cut off diplomatic discussions. The throngs of Chinese tourists who usually visited Japan every week started to dwindle. The conflict was escalating.[17]

On September 21, Japanese trading houses informed authorities at Japan's Ministry of Economy, Trade and Industry that China was refusing to fill Japanese companies' orders for rare earth elements, including neodymium and dysprosium. These metals—the essential materials in Japan's high-tech industry, well known for its high-quality components—were virtually all produced in China.

Beijing never acknowledged an export ban or said that it would use rare metals trade as a political weapon. But no other country reported such delays in receiving rare earth shipments. And the Chinese officials never explained why all thirty-two of the country's exporters of rare earth elements halted trade on the same day. Restricting rare earth exports was an astute move if Beijing's goal was to escalate the political conflict between the two countries without the use of force. Because Japan receives more than half of all Chinese rare earth exports, a supply cutoff was far more important to Japan than it was to any other country. Tokyo feared that a prolonged ban would have dire economic consequences for their companies. Officials worried that rare earths were just the beginning of what China might withhold because China is also the leading global producer of twenty-eight advanced metals also vital to Japanese industry.[18]

Bowing to Beijing's pressure, Tokyo released the Chinese captain a few days after China cut rare earth exports. But the damage to Japan and the rare earth market had just begun. Prices for rare earths started to climb, some as much as 2,000 percent over the next year and a half. The tensions revealed that despite Japan's economic prowess, the country was still susceptible to the restriction of a handful of metals few had heard of, as was the case three decades earlier when cobalt was in short supply. The events also laid bare the design shortcoming in Sagawa's magnet, which he told me was a great regret—relying on one of the least produced rare earth metals, dysprosium, to make permanent magnets.[19]

Decades after Sagawa's magnets first made their way out of the lab, it is now clear that rather than shoring up supply lines and increasing the country's economic security, his invention merely switched Japan's resource dependency from Congo to China and from cobalt to rare earth elements.

To understand the link between rare metals themselves and the products that consumers use, I set out to meet Sagawa. As I approach his house at 9:00 on a Saturday morning in July, I see him waiting. And although we have never met, he greets me with a warm smile, recognizing that this 6'3" Westerner, navigating the twisting roads in the foothills of Kyoto—an area in which he has lived for nearly thirty years—is his visitor. He guides me up a small path to his house and into his living room. He offers me a small plastic bottle of a pink berry juice as I sit down. My eyes dart around the room catching sight of a glass cabinet holding several awards, including the Japan Prize, the country's highest award for international scientists.

This seventy-year-old scientist has a waft of thick jet-black hair and projects a professorial air, but has the enthusiasm of someone decades younger. He speaks English with a heavy Japanese accent, though he doesn't miss a word. Sagawa tells me that I am only the second foreigner to interview him for his story, a sign that the world has overlooked his achievement. Indeed, he felt it was so important to explain his work that he scheduled our meeting just several hours before one of the most momentous events in his life: his only daughter's wedding.

Sagawa lays down a green handkerchief on the dining room table between us. He peels back the corners to reveal his invention: a two-centimeter square shiny metal block—a permanent magnet, made from iron, boron, and two rare earth elements, dysprosium and neodymium. And they reveal their powers quickly.

He tries to separate the metal block into two pieces, but it clasps back together, pinching the fleshy piece of his skin between his thumb and forefinger. Like a beekeeper who has just been stung, Sagawa laughs off the pain—a part of the job— but a dangerous one. His are no ordinary refrigerator magnets. They are more than forty times stronger. The pull of these magnets can break bones. They can hold more than a thousand times their own weight, the same strength a newborn baby would need to lift an elephant.

Sagawa tries again to separate them, placing one silver block on either side of his hand. He flips his hand back and forth—the magnets never move. He gives them to me to look over; it's not one block but approximately six individual shiny silver slices, each roughly a quarter centimeter thick. As I set the block down, it skitters across the table toward my iPhone. Sagawa looks concerned as I separate the two and asks if the phone is still working. Fortunately, it is. He tells me I am lucky because these magnets can erase memory from electronic devices within ten centimeters of them, which highlights not only their power but also the challenge of shipping them.

Staring at this silver block, I ponder its power. Just fractions of a gram of this rare earth magnetic material provide the spark that vibrates my phone. The rare earths are also the linchpin of its sound system. An electromagnet produces a magnetic field that changes in intensity to rapidly attract and repel a permanent magnet inside the speaker. This creates vibrations that the speaker helps to turn into audible sounds such as a ringtone or my mom's voice.

The beauty of Sagawa's rare earth magnet is that it efficiently converts electrical energy to motion in motors, which has allowed for the miniaturization of electronics, the creation of cleaner, greener motors and more accurate weapon systems. Its efficiency means that these magnets and the rare metals in them are nearly everywhere. They perform varied tasks that may appear to the layman to have nothing to do with magnetism, but they do. Rare earth magnets help computer hard drives to retain information, increase energy efficiency in air conditioners, and propel our hybrid vehicles. As Sagawa's invention plays a hidden but essential role in modern society, this makes him one of the most significant inventors of modern times. Yet, this unassuming man and his invention are known in only a limited circle of scientists. It turns out that Sagawa, like the rare metals and components he toys with, languishes in obscurity. But that changed in 2010.[20]

Japanese officials are now more concerned about these dependencies and a loss of manufacturing operations as companies leave for China to secure resources. After the scare of being cut off from Chinese rare metals, Tokyo gave Sagawa more than $10 million to remove dysprosium from his magnet. By 2013, Sagawa had reduced the amount of dysprosium in his magnet from 10 percent to roughly 3 percent, but getting the new magnet into the real world will require several more years of testing. Reducing the amount of dysprosium per magnet has yet to be a long-term solution to dysprosium shortages because the modest reductions have been dwarfed by the increasing number of magnets that the world needs.[21]

Ideally, Japan would love to create a dysprosium-free permanent magnet, but that's a goal that Sagawa and other researchers see as unlikely. Perfect substitutes for natural products are not easy to find. Iowa State University material scientist Karl Gschneidner—also known as "Mr. Rare Earth" because of his lifelong study of those elements—points out that the world has looked for alternatives to these permanent magnets for more than thirty years, but with little success.[22]

And of course, reducing dependency on Chinese rare earths won't solve the fundamental resource concerns that Japan and the rest of the world are facing. Just as rare earths replaced cobalt, any new scientific breakthrough to replace a minor metal in one application will merely shift demand to another one with its own political sensitivities and vulnerabilities. New technologies mean the world is bound to use more and more minor metals—after all, cobalt demand has tripled globally since the late 1970s—which in part makes these minor metals increasingly important in geopolitics.[23]

As the rift between China and Japan demonstrates, a country's resource security is not just about oil and gas anymore. Global resource needs have broadened, which raises the specter that minor metals will increasingly be at the heart of conflict between individual resource-rich and resource-poor countries, and will drag in manufacturers and producers. U.S. senator Duncan Hunter is very concerned about the geopolitical leverage that rare metal–rich countries can apply. They "may be able to exert enormous leverage over [a resource-poor] country, not by attacking those tanks and trucks and ships and planes, but by attacking the supply of his components."[24] It's a lesson that the United States learned well a century ago.

Before the onset of World War I, most countries paid little attention to the crucial role that resource supply lines played in battlefield success. But as the nature of fighting changed—becoming more mechanical and relying on heavy artillery and armaments rather than on the number of soldiers and their hand-to-hand combat skills—resources took their place on the front line. The new weapons required base metals like iron, but they also increasingly required rare and once obscure metals like tungsten and molybdenum to make weapons stronger and more heat resistant. During World War I, tungsten, a steel strengthener, became so central to the war effort that the United States deemed it treasonous to export it to enemy countries.[25]

After the war, a scarcity of many minor metals and other resource materials led Major General James Harbord to create a list of forty-two strategic materials, including the rare metals vanadium, tungsten, and chrome that the War Department deemed critical in war. Meanwhile, Germany, smarting from its loss, expanded its metallurgical capacity by refining copper and other metals, developing synthetic alternatives for materials, and signing long-term contracts with mining companies.[26]

As countries slipped into World War II, Charles Merrill, supervising engineer of the U.S. Bureau of Mines, warned, "In time of war, victory or defeat may hinge on the availability of strategic materials."[27] The United States decided it had to fight on the resource front as well. The government also beefed up domestic funding for research and development of mining sites and plant construction, and it bought domestic resources for the war effort and eventually stockpiled nearly fifty different minerals.

To thwart Japan's growing militarism in the late 1930s, U.S. president Franklin Delano Roosevelt stepped up a "moral embargo." He encouraged U.S. manufacturers to restrict exports of airplane components and a number of rare metals such as molybdenum, tungsten, and vanadium to Japan. As the war took shape, Washington pressured its Latin American neighbors not to sell their resources to enemies, and used U.S. oil exports as leverage to encourage Spain and Portugal to cease trading tungsten with Germany. At the same time, the United States spent about $2 billion to buy minor metals globally, often engaging in bidding wars with Germany that pushed the price of some materials ten to twenty times higher than their peacetime levels. By the end of the war, the Allies had been able to cut off the Nazis from chrome and tungsten, among other metals, a feat that, as a Defense Department official in the Reagan administration later commented, "halted the Nazi war machine."[28]

Having learned the importance of shoring up access to critical materials during the war, President Truman established the Material Policy Commission in 1951. It developed a framework to ensure that U.S. military businesses had the resources to compete in an increasingly global marketplace, which constituted one of the largest reports in the nation's history—if you stack the twelve volumes of the findings they extend over a foot from end to end. Internationally, U.S. diplomats served tours of duty in a newly created office of nonferrous materials policy.[29]

Interest in critical materials remained high in policy circles through the 1960s, and the oil shock of the 1970s sparked concern once again about critical resources. But fears and concerns over critical materials dampened with the recession of the early 1980s as it ushered in an era of relative price decline, and in the case of some commodities, like oil, far lower prices. Western governments started to feel that natural resource security, especially critical material supply, was yesterday's issue. Both U.S. president Ronald Reagan and U.K. prime minister Margaret Thatcher believed if markets were left alone, businesses would allocate resources appropriately. Despite resource shortages early in his term and concerns about Soviet control over critical materials, the Reagan administration rejected the need for governments to subsidize mineral production even for defense purposes.[30]

At the same time, U.S. government research into rare metal supply lines slowed; resource security policies stagnated; and in some cases, the government even purposefully weakened its safeguards. Between 1993 and 2005, the Defense Department sold more than 75 percent of its stockpile, underscoring the country's faith in commodity markets.

This drive to free markets was at the root of increased demand for natural resources. Millions of consumers in China and former Soviet states became unhinged from the constraints of a controlled market and started to consume like those in the Western world did. This laid the seeds for greater use of minor metals and with it, the price increases of the mid-1990s. This change marked the end of an uneasy price decline that had started decades earlier.[31]

With more global consumers and global manufacturing now relying on more rare metals, resource-rich countries have become keenly aware of the geopolitical and economic power

that rare metals afford. Countries such as Japan that are unable to meet resource demands either through their own mining and production or through secure stable trade, will find
companies leaving their shores and taking thousands of jobs
with them. When it comes to rare earth and other rare metals,
this is a migration that China is counting on.

The importance of rare metals and specifically rare earth elements in China's development goes back to the country's revolutionary leader, Deng Xiaoping. In 1992 he said, "There is
oil in the Middle East; there is rare earth in China." By then,
China, out of necessity, had begun to mine its rare metal resource deposits. Fifteen years earlier, the country had started
developing its manufacturing and construction sectors, which
led to a growing reliance on imported material supplies. To
reduce that dependence, China increased its investment in
primary products it found domestically—including mined
commodities, mostly minor metals.[32]

China was often able to produce many rare metals at a
50–60 percent discount due to lower labor costs and environmental standards. This quickly opened international markets
and produced hard currency from exports as well as stable
metal supplies for domestic manufacturers. To increase exports, China introduced value-added tax rebates to its exporters to further undercut foreign prices. Companies outside
China could not compete with the lower costs. Tungsten mines
in Germany and France shuttered in the 1980s, likewise rare
earth producers in the United States in the 1990s. Soon manufacturers came to rely on those low prices as the new normal.
And China became the only seller of many of these critical
metals including rare earths.[33]

Beijing's policies soon became more sophisticated. They realized that selling ore abroad was a lost opportunity. With near monopolistic control over most rare earth mining, Chinese industrial policy promoted the development of making material and components from rare earth elements. Over time, the rare earth supply chain was at the heart of a strategy to build high-tech companies that create jobs at home. Instead of exporting rare earths to help create jobs in Japan, China wanted to build high-tech manufacturing plants to use its own resources and employ its own citizens.

To achieve these goals, China reversed its export incentives for rare earths and other metals during the beginning of the first decade of the 2000s and began export restrictions, including quotas, to stem the overseas flow of resources. Lower domestic prices enticed foreign companies to bring their operations to China for unrestricted access to its abundant rare metal resources base. Echoing Deng Xiaoping, Gan Yong, the head of the China Society of Rare Earths put into words what had long been Chinese policy in 2013: "The real value of rare earths is realized in the final product."[34]

China's unabashed attempt to control the entire high-tech supply chain, from rare earths to finished products, worries many. But no one has outlined the risks quite the way Gerald van den Boogaart has.

If one were to call central casting for a German mathematician, it would likely send you Boogaart from the Helmholtz Institute Freiberg for Resource Technology. He has salt-and-pepper black hair with floppy black bangs that extend unevenly to his silver wire-rimmed glasses. He speaks with a strong German accent, and appears slightly frazzled, although he conveys an air of confidence.

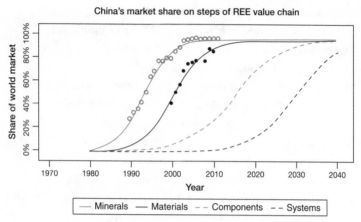

Figure 2. China's market share of global production of various
rare earth elements (REE) of increasing value.
Source: Karl Gerald van den Boogaart, Polina Klossek, and Andreas
Klossek, "How Forward Integration along the Rare Earth Value Chain
Threatens the Global Economy," paper presented at the 2014 Critical
Minerals Conference, Denver, Colorado, August 3–5, 2014. Data
based on Roskill Information Services and Brose Group.

As a mathematician, not a rare metal expert, he makes
for an unlikely speaker at the Society for Mining, Metallurgy
and Exploration's 2014 Critical Minerals Conference in Denver. In a presentation titled, "How Forward Integration along
the Rare Earth Value Chain Threatens the Global Economy,"
Boogaart argues that China will soon dominate the production of high-tech goods from computers to trains as well as
their components. At the heart of his argument is a graph
containing four parallel upward-sloping s-curves. The first
two s-curves of Figure 2 track the historical progression of
China's share of global rare earth mining from the late 1980s,
when the country produced a fraction of the world's rare earths,
to after the turn of the century, when China monopolized pro-

duction. The second line followed a similar trajectory several years later for China's rare earth material production, which also increased.[35]

The next two s-curves are Boogaart's ideas of the future; one demonstrates China's increasing global share of rare earth component manufacturing, and the other, products that will rely on them, including cars, wind turbines, and magnetic resonance imaging machines.

From a strictly economic standpoint, Boogaart is concerned: while the rare earth industry, in his estimation, produced revenues of $4 billion annually, the materials, high-tech component, and user systems industries were each larger, by a factor of ten—meaning that the production of these systems is worth about $4 trillion in total. For him the Chinese authorities have a simple calculus: If they could reap trillions in the future by investing billions now, why wouldn't they do it?

He predicts that over the next several decades, every high-tech system—from cars to solar panels—could very well be produced in China, thus sucking up the manufacturing base of other countries, just as it has done with the production of rare metals.

A shift is already under way. Over the past decade, industries in Japan, which were the biggest export market for Chinese rare earths, quietly moved operations that rely on rare earths to China. Michael Silver, chief executive officer of the California-based American Elements, tells me, "We have witnessed in my business a huge number of our customers finally building facilities in China to access elements."[36] He adds that historically the United States supplied all of the cerium and neodymium for General Electric's lightbulbs. "All of that is done in China now. Similarly, every major wind turbine manufacturer has moved operations over to China."[37] Japanese officials relay similar stories.

It's not a story that China is hiding. In 2011, when China's vice premier Li Keqiang went to Japan, he asked the Japanese to bring their technologies to his country. "Japan has advanced technology to develop and use rare earths," Li said. "I hope both countries' companies can promote cooperation."[38]

China is not unique in using its own resources for its own development—the United States banned the export of certain petroleum products, and other resource-rich countries including Indonesia, Chad, and Zimbabwe restricted exports. But China is the dominant rare metal player, so its policies raise global concerns.[39] The real problem for China is that their export controls were illegal, at least according to the World Trade Organization (WTO). The export restrictions create an unfair advantage for domestic producers and, as the European Union claimed, "put pressure on foreign producers to move their operations and technologies to China." For example, in April 2014, rare earths prices were 36 percent lower in China than overseas. The United States, Japan, and the European Union pressed their case at the WTO in 2012 and won roughly two years later, handing China its second loss involving rare metal policy at the World Trade Organization in less than three years.[40]

Many economists believe that the battle for rare earth elements is over. China lost the case and adjusted trade policies while rare earth prices are well off their peaks. But this war over the periodic table is a long-term struggle, and it is far more likely that any of China's losses at the WTO will only change the country's policies, not its strategy. Since China controls an increasing percentage of rare metal production—about 40 percent of all minor metals in 2010, up from 29 percent in

2000—the risk remains for those who rely on the rare earth elements and, indeed, for all rare metals that China produces.[41] Even if companies and countries shift away from certain rare metals, as the world once did with cobalt, they are still setting themselves up to relearn the lesson that substitution might not make them any more resource secure, as an underlying instability in the market will remain.

The real challenge for our resource future may be more than geopolitical battles. Because even if a country is willing to trade, other obstacles to ensuring rare metal supply will be far greater.

III

Corporate Hurdles
Monopolies and Investment Incentives

For centuries, the small Brazilian town of Araxá welcomed visitors to explore the region's green rolling hills and thermal, but slightly radioactive, spas. A few miles from the city center, just below some of the most lush, verdant terrain outside of a rain forest, sits the $300-a-night Hotel e Termas de Araxá—a gem of colonial-style architecture, built by a former Brazilian president in the 1940s to attract a well-to-do domestic and international clientele. The complex is stately; with the exception of government state houses, few places boast a three-story rotunda with a stained-glass ceiling and a white and black marble floor. But for all its grandeur, the resort's free-wheeling casino days are behind it. Seventy years after its founding it has lost some of its luster: grass rises up between the cracks in the cement-blocked sidewalk. I doubt if more than 15 of the 283 rooms were occupied when I visited in 2013.[1]

Other than the hotel, Araxá has lost much of its tourist appeal. A woman from the region's capital city, Belo Horizonte, who befriended me when I had difficulty checking into the hotel, complained about the lack of sophisticated nightlife. She told me she prefers to stay in her room when she comes here on business every other week.

The truth is that tourism, despite a longer, more romantic hold on the region, lost top billing to mining decades ago. As a local geologist told me when I asked what was more valuable, the hotel or the minerals beneath, "the minerals, of course." Above the resort are terraced, carved, unnaturally beautiful hills. Their lines gracefully cut through the hillsides providing a life-size topographical map of hills long gone, sheared off during the mining process over decades. The grandeur of the hotel now looks misplaced. Despite the inherent contradiction of tourism and mining existing side by side, they are both in Araxá for the same reason: a caldera.

A former volcano that produced the town's hot springs also left a mineral-rich legacy. Magma seeping through the continental crust formed a mineral complex rich in niobium, rare earth elements, and phosphate. Since 1955, Araxá has been home to Companhia Brasileira de Metalurgia e Mineração (CBMM), the world's largest miner of the soft, granite-gray metal niobium, and to its smaller-producing neighbor, Vale Fertilizantes, a subsidiary of the mining giant Vale, which mines phosphates for use in fertilizers. Araxá is also one of the few places I have visited where ecotourism and mining seem to coexist in harmony. Many times the relationship between environmentalists and miners can be, at best, testy.

Across the parking lot from the resort is a well-built, cream-and-pink stucco building of five stories. Despite the "Hotel Colombo" sign above the entrance, there are no rooms to reserve. With dwindling tourists, the town turned the hotel into a city office a few years back. The government, seeking to capitalize on the region's wealth of resources, plans to use the building as a technological hub to attract research scientists and manufacturers. A forward-thinking mayor wants to double down on the region's mining future, and few in the tourist town seem to protest openly. The town's employment and economic base is now built on the mines, not the hotel—the town earns about 80 percent of its income from mining. CBMM is fortunate; the company faces little acrimony.[2]

However, elsewhere mining faces greater opposition. Even in Colorado mining country, where tourists now visit old mines, opening new mines, especially radioactive ones, presents great obstacles. CBMM is fortunate that its mine is older than most people in the town and grew with it. It's a time advantage that new mines lack. Addressing environmental concerns is just one of the many hurdles new mining companies face. They also must navigate government regulations, unforgiving climates, and stingy financial markets. But it is sometimes the most basic hurdle that stumps them: you can only mine where there are resources to dig up.[3]

Although supplies of most minor metals exist in the world's crust to meet demand for hundreds of years, they are rarely found in high enough concentrations to be mined profitably. Like oil, some regions are blessed with rich concentrations, which gives them influence over the market. Oil analysts point out that Saudi Arabia's position, with 16 percent of reserves and a bit over 10 percent of global production, is so dominant that the country historically has had control over oil prices. But most production concentrations in the rare metal sector are far higher and therefore so is their control. For example, the local government in Lengshuijiang in China's Hunan Province controls about 60 percent of the world's antimony production.[4]

Relying on one country and one mine in particular is a risky proposition. A natural disaster, political changes, or conflict such as we have seen in Congo can quickly create shortages. Concentration of supply in one mine is why some in the industry are concerned about CBMM's dominance in the market. It produces 85 percent of the world's niobium. In fact, the mine is so critical to the global trading system that the United States places CBMM's Araxá mine on its list of critical infrastructure abroad.[5]

While CBMM is one of the most upstanding and well-run mining companies I have seen and has certifications to back up its efficiencies, I sensed that a level of secrecy pervades the company. CBMM employees whom I e-mailed or called did not respond or they referred me to one man, the chief executive officer (CEO) Tadeu Carneiro. There are incentives to be loyal. Their children receive education subsidies through their university years; the firm covers medical and housing costs; and the pay is great, they tell me. Out of 1,800 employees fewer than 10 leave in a given year.[6]

In 1985, the National Materials Advisory Board of the U.S. National Research Council summed up the company's market position, "CBMM not only has the largest capacity . . . but by far the lowest costs. Thus, CBMM could at any time cut prices and effectively drive its competitors out of business. However, its current strategy is to let its competitors survive and to integrate downstream."[7] Not much has changed.

What's good for CBMM is not necessarily good for the market. The company produces niobium from a reserve that can meet the world's needs for two hundred years and is efficient and extremely profitable. The company's owner, the Moreira Salles family, has an estimated $27 billion in assets, around half of which comes from this mine. According to

Bloomberg News, the mine runs a 37 percent net-income profit margin and earns more than $600 million in profits annually.[8]

Some consumers feel that the market could benefit from more competition. CBMM had become so successful that its customers decided to become owners. Concerned with the reliability of their supplies, Japanese and Korean companies, some of CBMM's biggest customers, bought a combined 15 percent stake in the firm in March 2011 for $1.8 billion, with support from their governments. China followed suit acquiring a 15 percent share six months later for $1.95 billion. But even with ownership, the Asian minority shareholders were forbidden to carry out technical due diligence research, *Bloomberg News* reported. You sense the level of secrecy once you arrive at the mine's main gate.[9]

Even with the mayor of Araxá, Jeová Moreira da Costa, in the car with me, my car has to pull into the parking lot next to the CBMM guardhouse. A neatly dressed guard asks for my passport. He dutifully checks it along with the other passengers' identifications. I look back at the main gate and see another security guard searching through the trunk of a car waiting on the other side of a closed exit, likely looking for items that shouldn't leave the mine.

As we enter the site I'm struck by the incongruity of the immaculately trimmed grounds, a cross between a leafy country club and an industrial technology park, and the area of stripped earth in the distance. It's hardly a stereotypical mine: palm-tree-lined streets, green manicured lawns, and matching khaki and blue buildings, some with smokestacks. As I admire the identical jumpsuits of the workers waiting at a bus stop inside the mine site, the mayor leans over and says, "The city is happy for a beautiful company."

We arrive at a visitor center where a spotless white door-mat greets me. As I set my footprint on it, I look up to see Tadeu Carneiro, CBMM's CEO. He is sporting blue jeans and a blue button-down with a lapel pin of his alma mater, the University of Pittsburgh. Carneiro has dark black hair, prominent eyebrows, and a gravelly voice. He is innately charming; his Portuguese-tinged English, his measured speaking pace, and a penchant for slipping your name into his speech convey warmth and connection.

Shortly into his introduction to the mine he declares, "We're a technology company, David, no doubt about it." The phrase is a cliché that mining executives use to change the focus on the company from the digging itself to the applications that the digging helps to create. Carneiro explains that the mining is just one of the company's fifteen technological steps. Turning rock into metal, or at least a material rich in metal that can be used in today's products, is a technological feat separate from mining.

CBMM is fortunate in that their soil is loose and they don't need to use the most expensive blasting procedures to break up a rocky base. Its mine is also aboveground, which makes production far cheaper than digging tens or hundreds of meters beneath the surface. CBMM uses a combination of trucks and a conveyor belt to dig up and move the ore containing 2.5 percent niobium from the mine to the processing facility. And because of the loose soils, they also don't need to spend as much as other miners to grind up and sift the ore. (Crushing and grinding consumes nearly half the energy used at a mine, so less grinding means higher profits.)[10]

CBMM then removes waste—anything that does not contain niobium—from the ore with the help of filters and magnets and eventually water, heat, and acid. The goal is to further isolate the minerals it wants by separating out the ones it doesn't.[11] After the niobium-rich ore is crushed, it is placed in a bubbling, oily acid brew. The niobium material sticks to the bubbles, which separates it from the rest of the waste. CBMM's refining produces a concentrated niobium material that is about 50–60 percent niobium. The company then refines it further, mixing it to meet the demands of customers, most of whom are steel processors. Adding just a pinch of niobium per ton can reduce the amount of steel needed in its most popular applications: pipelines, bridges, and turbines. When Gustave Eiffel started constructing the tower that would eventually bear his name, he needed 7,000 tons of steel; today if you wanted to build a replica Eiffel Tower, you would only need 2,000 tons of steel because of CBMM's niobium.[12]

"Forty years ago, niobium was just a theoretical possibility, a dream in a laboratory," Carneiro tells me. Based on his rhythmic intonations this seems to be a story he has told many times before. As I listen to him speak about the development of the niobium market, he could have been speaking about the development of the market for nearly all rare metals over the past sixty years—from dysprosium to titanium. They once served little purpose, but as new uses have developed, the metals have become irreplaceable.

In the 1950s, when CBMM started operations, niobium had little market use although research from Britain and the United States had discovered that it strengthens steel, while simultaneously lightening it and raising its heat resistance. These are important attributes because British steel warships literally cracked in the cold waters of the Atlantic during World War II and three airplanes broke apart in flight.[13] Material science had not caught up with the dreams of naval and aerospace engineers of the time. Unable to develop profitable mining locations and having a difficult extraction process, few focused on researching how to use niobium.[14] Enter CBMM.

Since CBMM was selling a material that no one knew they needed, unlike copper and iron, the company had to create a market. Its sales strategy was to increase demand by finding uses for niobium while simultaneously improving metallurgical processes to drive prices down to convince skeptical customers to try it out. By the 1960s, CBMM had lowered production costs, leading to a fourfold drop in prices.[15]

Beginning in the same decade, company research led to advancements that significantly increased steel's toughness and strength by refining the grain structure in the niobium-infused steel. By the 1970s, steel manufacturers began taking note of CBMM's advancements. During that decade, CBMM grew almost 10 percent annually. CBMM was shaping the steel-making market, and its material found its way into buildings and pipelines. In the 1970s and 1980s, niobium seeped into more construction materials, shipbuilding, and offshore oil-rigs. Carmakers also used more niobium in an effort to build lighter cars and thus to reduce fuel use.[16]

Seeking greater sales, the company researched ways to increase the amount of niobium per ton of steel. It was a challenge. Raising the concentration of niobium in liquid steel makes the liquid thicker and pulpier. And as would happen if one added too much flour to a pancake batter, it wouldn't pour easily. The mixture wouldn't produce high-quality steel. Over time, CBMM scientists unlocked the secret of alloying steel with more niobium and sales took off, increasing by a factor of four between 1990 and 2010.[17]

Twenty-five years ago Brazil, the leading consumer of niobium per ton of steel, used 40 grams per ton. Now 40-grams-per-ton steel—the amount that China uses in its domestic market—is considered low quality. High-grade steel can absorb 100 grams per ton and, with it, achieve higher levels of strength. For specialty steel, such as pipeline steel, CBMM has found ways to use even more.[18]

"We never worried about fighting our direct competitors," Carneiro says, referring to CBMM's growth period. This is perhaps understandable when one has established a dominant market position with some of the lowest-cost resources. "If people want to develop them," referring to the roughly three hundred other similar niobium deposits around the world, "that's fine." But Carneiro makes a point that might be disheartening to potential challengers, "we will continue to increase manufacturing processes here."

This certainty and security that it can keep digging gives CBMM an undeniable edge in its industry. While future technologies can help make the excavation and processing of ores cheaper, mining companies without politically secure, technologically feasible, and geographically accessible deposits like Araxá's niobium must now go to more and more remote locations and dig deeper and deeper. Most of the easy-to-access, high-grade, inexpensive mineral deposits have been mined.

For many smaller, struggling mining companies, CBMM and its success exists in a parallel universe. To understand the challenges that new mining companies must confront, look no further than Tiomin Resources. In 1995, J. C. Potvin, CEO of the Canadian-based firm, thought he had found a goldmine, so to speak, a rich titanium-producing deposit, forty kilometers south of Kenya's port city of Mombasa.[19]

Potvin began the process of securing government licenses and geological information, hoping to begin commercial development in 1999. Over the next four years, his company completed a feasibility study, a company-funded report to determine the economic viability of the project, as well as the first of what were to be many environmental assessments. Potvin was convinced that the project would be successful. Despite boasting of "excellent cooperation and support" from the Kenya government, operating in Kenya, as in many developing countries, is not straightforward.[20]

Environmentalists slowed the project because of fear that mine pollution would upset the surrounding lands, and farmers complained that the compensation the company offered for their land was too low. The Kenyan government, which initially tried to take an ownership stake in the mine, soured on the deal after the 2002 election brought in a new government with a more cautious view of the project. For the next several years, the government, the courts, and the company could not come to terms about resettling approximately 375 families who had to leave their land or about addressing environmental concerns.[21]

After almost a decade of wrangling, Tiomin signed a twenty-one-year land lease with the government in 2004 and worked out a royalty payment scheme. By then, the price of the stalled project had begun to creep up from an initial $120 million to nearly $200 million while the legal battles continued. On December 19, 2006, the court finally ruled on the last court cases in Tiomin's favor; its stock soared 24 percent as the company appeared set to begin construction. But the market was overly optimistic because Tiomin's plans quickly fell apart. The company could not remove all the villagers from the land. Seven families held out, demanding fifty times the land value.[22]

The company's big investors—wanting government assurances that the project would go forward—soon backed out. Election violence in 2007 and 2008 made working with the government impossible because authorities were focused elsewhere. In 2010, an exasperated company board sold out to Australia's Base Resources. With greater financial backing, government stability, and better timing, Base completed the mine at a cost of $305 million. It was producing for its first commercial sale in 2014, close to two decades after initial exploration.[23]

Despite the challenges Tiomin faced in Kenya, our high-tech future relies on mining in far more chaotic places. Resources supply lines are stretching longer not only through more inhospitable environmental terrains but also increasingly through areas fraught with political turmoil. In Colombia, FARC rebels, who have been fighting an insurgency against the government since 1987, produce tungsten from the depths of the Amazon jungle. In Democratic Republic of Congo, anti-government forces and rebel gangs make millions producing tungsten, tin, and tantalum. In 2011, about 21 percent of the world's tantalum supply came from regions in conflict, and almost all of it was processed in China. On the twin Indonesian islands of Bangka and Belitung, bands of small-scale illegal miners dig up more than a third of the world's tin from jet-black cassiterite minerals, and unknown amounts of other minerals like xenotime and monazite, which hold rare earth elements.[24]

Countries that are growing wealthy are starting to pose more challenges to metal operations because as their citizens become wealthier and buy houses with backyards, they are less willing to accept pollution in them. These citizens have more time and resources to fight mining projects and they want just compensation for any environmental damage. In 2012, in Malaysia alone, citizens were fighting against three large mining-related projects including a "light" rare earth processing facility built specifically to reduce the world's reliance on Chinese production. Meanwhile, demonstrations in Sichuan Province in China shuttered a proposed molybdenum processing facility on the heels of similar protests in Dalian and Xiamen.[25]

The concern now—due to the paucity of investment since the financial crisis and the long lead time needed to develop rare metal projects—is that sooner or later, our visions of a high-tech future will collide with the reality that we didn't invest enough in ensuring an adequate supply of critical materials. The challenge is that the pace of innovation and therefore our demand for rare metals will increase at a far faster pace than those planning and building mines can supply.

In this sense the production of rare metals from scratch is akin to the distilling of Johnnie Walker Scotch Whisky or fine cheddar. You need a long lead time of years from when you start producing to when you can enjoy them. Rare metal mines (and supply chains) can take up to fifteen years from investment to production according to the U.S. government and as experiences like Tiomin's show. That means the rare metals coming out of some mines today were made in 2000. That is a lifetime ago for the tech field where many of these metals go. It's hard to imagine that the executives who were producing material for 2015 had any concept of the smartphone or many of the products their materials are used in. At the time, they were focused on the potential fallout from the Y2K bug. Today, we are asking for tremendous technological foresight of our mining executives, especially those of junior mining companies, because they identify and evaluate new mineral deposits.

They are the ones who should know where their materials will be sold in fifteen years.[26]

The challenge is that junior mining companies are like biotech start-ups; most, even those with the most promising projects, will fail. But without biotech, the next great cancer drug won't be developed; without these junior miners, the world won't have the ingredients for the next green power source. And just like biotech companies, these companies won't get anywhere without investors.

Just before 8:30 on an early Sunday morning in April 2013, Ron MacDonald, the plenary speaker at the InvestorIntel Technology Metals Summit, walks up to a three-foot-high stage at the Sheraton Centre Toronto Hotel conference hall. MacDonald, the former Canadian secretary of international trade, is now an executive at a vanadium mining firm and a rare metal evangelist. In this role, he faced a formidable challenge before a crowd of mostly rare earth exploration executives. His goal was to raise the spirits of junior mining executives who have watched the price of rare earths (along with their companies' share prices) plummet by 80–90 percent over the previous eighteen months.[27]

Funding for mining companies had been hard to find since the financial crisis, and in 2013, the situation was getting worse. For the first time in almost ten years, there were no mining IPOs (initial public offerings) on the mining-heavy Toronto Stock Exchange during the first quarter of 2013. What little money was coming into the sector was going to larger companies the size of CBMM, not to the people in this room, who have spent their careers finding new rare metal deposits.[28]

During the first three months of the year, junior mining companies raised less than 3 percent of all mining funding; that's less than $102,000 for each junior company globally.[29] Because most of these companies don't yet sell metals, they survive on investment alone. No investment means no jobs. And the future looked even more dire.

Before the Toronto conference, IntierraRMG, a leading data provider to mining companies, pronounced, "The industry is in for a bleak year." The consultancy PwC (PricewaterhouseCoopers) was even less sanguine, blunting stating, "You have a perfect storm for the demise of many entities which play a critical role in exploration. . . . This will have a dramatic impact on the pipeline of new reserves." For the world to develop reliable sources of critical materials, these companies needed money, today. Some exploration companies were so desperate that they took to Twitter, LinkedIn, and Facebook to troll for dollars. The mood was so gloomy that conference organizers had even scheduled a lunchtime motivational speaker.[30]

As the opening speaker on the first day of the conference, MacDonald picks up on the room's downbeat vibe and is determined to turn it around. He speaks with conviction and in catchphrases: "Forget what's going on in the market." "We are feeling a new wave." "We are talking about a trillion-dollar-a-year industry that is going to come out of the doldrums!" But he must have also noticed the key problem I saw. Although I spotted several consultants, a government official, and industry analysts in addition to junior mining company executives, the investor conference was awfully short on its most essential participants: investors.

Instead of an air of excitement about deal making, the event had a Thanksgiving-at-the-in-laws vibe. Everyone knew everyone else's business and they all appeared to be attending out of obligation rather than desire.

Even in the best of economic times, rare metals can be a tough sell. Take MacDonald's vanadium mine. Before he can convince investors to hand over the cash, he has to explain—over and over again—what vanadium is. It's hard to give an elevator pitch for something no one has heard of.

MacDonald could boast about vanadium's primary application: by adding less than 0.2 percent of vanadium to steel, steel's strength doubles and its weight falls by 30 percent.[31] (Vanadium is excellent for strengthening construction tools, where niobium is better for oil pipeline construction. Sometimes they are used together.) He could also talk about vanadium's crucial role in our society. In 1905, Henry Ford discovered that French carmakers used a vanadium steel alloy that was lighter, harder, and stronger than what he found in America. He imported the technology and used the alloy in the gears, axles, and shafts of the fifteen million Model T cars he produced. The material was so critical that Ford declared, "But for vanadium there would be no automobiles!"[32]

For MacDonald, all that's history. Like Carneiro, he's focused on his metal's future, which he tends to speak of in hyperbolic terms. MacDonald's company, American Vanadium, refers to its material as "The Holy Grail" because of another potential use: high-capacity battery-storage systems. MacDonald says that by storing power produced from intermittent energy sources such as wind and the sun, his vanadium power packs will become the central component in a global green power network.[33]

Running an early stage mining company is more often about salesmanship—convincing others to invest in your ideas—than geology. Executives spend all their time looking for resources, but more often they're financial rather than mineral. Even those like MacDonald who have been more successful in securing investment are facing far more obstacles than previous generations of mining executives did: increasing environmental regulations, the higher cost of mining operations and royalties, and more complex geologic formations. These all lead to greater costs. And as these costs rise, investors grow skittish.

But investors are right to be skittish: investment in critical materials is a massive bet. Is the resource the company plans to mine actually in the ground? Can the company develop the processes to separate the metals from the ore? Can material scientists find new uses for the extracted resources? And what metals will component manufacturers select to put in their new products?

While junior mining companies speak of becoming the next producing mine, most have smaller aspirations—such as raising enough money to get to the next stage of development. Assessing the geologic formation of a site, receiving the necessary government approvals, and proving the capability to produce a reliable product are preconstruction tasks that can cost tens of millions of dollars. If a company can show investors that it has a viable mine to bring to production, a high bar indeed, the company needs to find deeper-pocketed investors and banks to provide the hundreds of millions or billions of dollars—depending on the size of the resource—to construct it, buy processing equipment, build roads, and lay power lines.

Raising money as a public company means close market scrutiny and distractions. CEOs must field countless investor and analyst queries, time that comes at the expense of developing the mineral deposit. Such distractions are nuisances that

Carneiro at CBMM, a private company, does not have to worry about, admitting that setting up CBMM as a public company today would be far more challenging if he had to answer a barrage of investor calls.[34]

What's more, given the long lead time needed to establish mines, mining companies are trying to predict future demand to determine if there is even a market for their resource. Conversely CBMM had the rare good fortune of long-term support to invest in research and to find new uses. Establishing a mine with only a handful of scientists and engineers while constantly raising funds is a daunting task. Developing the equivalent of a CBMM today is a tremendous challenge, especially for public companies. The reality is that many junior exploration outfits are still trying to figure out what they even have or how they can get it out of the ground.

It's far harder for a new mine to enter the market against an established mine than it is for an established mine to produce more. But for resilient supply lines and cost-effective resources new mines and greater production are essential. Many of these upstarts, of course, want to reach the levels of success found at CBMM. But, if the past is any indication, most will either end up like Tiomin, having to abandon their dream, or face challenges like those of Avalon Rare Metal in northern Canada—caught with a plan, but not enough money.

Avalon Rare Metal's Thor Lake mine is emblematic of the high costs of starting and maintaining production. A five-hour flight or nearly five thousand kilometers' drive northwest of Toronto, the lake sits about five hundred kilometers shy of the Arctic Circle, adding new layers of complexity and cost because digging through tundra and refining minerals with water in freezing temperatures is more time consuming. Between 2000 and 2011 mining costs outpaced general inflation, nearly doubling, in large part due to increasing energy costs. And the costs have continued to rise. But their problem is representative of rare earth projects as well as mining in general, because the days of easy mining are over. Ore grades have been dropping steadily over the past one hundred years, and in the case of some metals, like copper, by 30 percent from 2001 to 2011. In addition, according to Pat Taylor, a chemical metallurgist at Colorado School of Mines, many newer deposits contain "problematic elements" such as arsenic or radioactive uranium. Finding a way to safely dispose of them adds greater complexity and cost. Today's new projects need more expensive infrastructure, which increases the costs of starting projects, creates higher operating costs, and ultimately leads to the production of more expensive metal.[35]

Avalon's extensive mining proposal is costly, and increasing. The company plans to dig up rare earth minerals on one side of a pristine lake, and then ship the ore containing traces of radioactive elements to a processing facility on the other shore. From there, Avalon will send the minerals roughly 1,500 miles to Louisiana for final refining. Even though production is years away, the price tag keeps increasing: in 2009, it stood at $400 million. Just a few years later it jumped to $1.5 billion. Fortunately for Don Bubar, Avalon's CEO, if its current rare earth project doesn't work out, his company has other locations to mine. As with most prudent junior mining companies, Avalon doesn't focus solely on one property.[36]

In 1979, when the first miners explored around Thor Lake in Canada's Northwest Territories they found a deposit they hoped would produce beryllium. Later developers hoped it would produce niobium. But due to its remoteness, challenging geology, and low market prices for those particular resources, no previous company had successfully developed the site. Bubar is trying to change that.[37]

But to be sure Avalon, which exemplifies what start-ups face, is not alone in initial low cost estimates and rosy outlooks. No doubt it is hard for CEOs to keep investors around for the long term, especially with countless cost overruns and delays.

Many rare earth companies are struggling with bankruptcy even though rare earth metal prices are higher now than they were when they estimated their revenue. The companies in production should be profitable based on their own previous assumptions. Many rare earth exploration companies as late as 2011 expected to raise hundreds of millions of dollars or more and to be in production by 2015, but those same companies are still years away.[38] Some always will be.

"Most of the executives are mining the market," says Jeff Phillips, president of Global Market Development and himself a successful minor-metals investor.[39] Part of the reason is that instead of determining which metals their deposit is best suited to produce, they identify the "hot" metal in the market—that is, the one that is most likely to attract funding to keep the company in operation. For example, Quantum Rare Earth Developments in Nebraska became NioCorp when its owners decided to mine niobium.

Dudley Kingsnorth, a former mining executive and the sought-after expert for rare earth elements, told me that a chairman of a vermiculite mining project referred to his deposit as an "Aladdin's cave" from which he would pick whichever metal in the ground he thought would get investors to give him funds.[40]

To understand why companies can develop mines for different metals on the same land, he says, you need to understand the mineralogy. Mineral deposits, like those at Thor Lake, were formed over millions of years. Magma, a liquid created by the melting of the planet's mantle and crust, rose from beneath the earth's surface, which allowed minerals to form. The unique composition of the magma and the conditions each pool of liquid encountered determined which metals formed— where on the earth's surface the magma arose, how long it remained hot, at what depth and under what pressure, and what substances it encountered along the way such as carbon dioxide or water.[41] Resource-rich areas often have a panoply of minerals; so where you find rare earth elements, for example, you will also find thorium, iron, or niobium. And since mining companies face expenses to develop each metal, they often select the potentially most lucrative ones.

There is an incentive to be optimistic; CEOs face a simple financial calculus that encourages them to keep every project alive. They can return money to investors and spend years developing a new site or keep at the current project even if prospects are poor. Returning the cash rarely happens. Faced with such challenges, some mining companies can take their salesmanship too far. Some use guarded language to describe their efforts in a positive light.

In 2013, Avalon stated in a press release that "an optimized hydrometallurgical process has been identified to crack all the minerals in the flotation concentrate with potential to improve heavy rare earth recoveries to over 90 percent along with improved zirconium, tantalum and niobium recoveries." My interpretation: company scientists have an idea of how to extract many rare earths from the ore along with other minerals, but they still have a lot of work to do to figure out if that idea works.[42] Relying on unclear industry jargon, many investors are often in the dark regarding the risks.

To reassure investors during this long process, many CEOs frame most news as good news. After all, the executives are all chasing the same limited funds and trying to stay one step ahead of the competition. Relentlessly positive sentiments make it challenging to distinguish the companies with realistic opportunities of success from those that are just struggling to stay afloat. The problem is partly of the investors own making because they are constantly clamoring for good news to raise the company's stock price.

Often the first casualty of this process is objectivity. "You need to separate the optimistic from the devious," John Sykes, an industry researcher at Australia's Curtin University, tells me. But that is hard for the average investor. Some companies play down the environmental, metallurgical, or economic challenges to bring a mine to production and play up the prices they plan to charge for their finished product. Company releases are subject to biases. According to Graham Lumley, author of the study "Mine Planners Lie with Numbers," quips, "If you do find objectivity in mine developments, can you see if my pet unicorn is there as well?"[43]

Studies, including Lumley's, argue that mining companies overpromise and under deliver; projects exceed their cost estimates and take longer than projected. Ernst & Young found that of the companies that actually reported cost overruns publicly, the average projects went over budget by an average of 70 percent. Lumley found that returns are 80–90 percent lower than company estimates. The mining giant, BHP Billiton, found that of the amount of steel that companies promised in 2008 to ship in two years, less than half actually reached the market.[44]

Although outright fraud is very rare, junior mining companies have a history of playing fast and loose with their business plans, a practice that has seeded a legacy of mistrust. In the mid-1990s, Bre-X, a small exploration firm founded by David Walsh in 1987 and initially run out of his basement, claimed to have access to some of the largest gold deposits in the world. The value of the company soared from \$.30 a share to \$250 in just a few years, before investors realized that the deposits were a fraud. After the story broke, Toronto Stock Exchange president Rowland Fleming commented, "If someone is simply lying, in a huge and stupendous fashion, as appears to be the case with Bre-X, all the disclosure rules in the world will not protect either the gullible investor or, it seems, even the most sophisticated one."[45]

A few months after the conference in Toronto where Ron MacDonald was speaking, I am at another investor conference in New York City, five blocks from my old firm, Lehman Brothers, where I was an analyst. As the moribund investment market lingers, the rhetoric continues to soar as the executives boast about the success of products that use their metals and the potential demand for them that this implies. The graphite producer raves about the Tesla electric car, winner of the 2013 Motor Trend Car of the Year; the rare earth miner points to advanced missile systems; and the vanadium producer heralds his metal's great promise in storing solar power. Instead of finding Hugo Boss–wearing, well-groomed investment bankers, most attendees seemed combination of older hangers-on hoping for one last-ditch effort to make a name for themselves and a younger generation of thirty-two-year-olds hoping to turn a \$30,000 investment into an early retirement nest egg.

One late evening after a conference in China, I sit down with a couple of industry analysts to talk about the individual investor who backs these junior mining companies. We decide it's a wide variety of folks, but there are a few main groups. To protect the reputation of investors I use caricatures: the German dentist, the Glenn Beck devotee, the weekend gambler. They all have visions of striking it rich, but their reasons for investing are different. The "German dentist," one analyst explains, just finished reading in *The Economist* how rare earth elements are crucial to numerous green-tech applications. He knows of a few rare earth elements such as thulium and holmium because they are in his dental tools.[46] After some research, he grows enamored with a specific mining firm's leadership and passion. He has a habit of repeating the same mantra that CEOs like Carneiro of CBMM voice: "These are technology companies, not mining companies."

We call another the "Glenn Beck devotee," who is mostly a gold investor, but dabbles in other commodities. He feels that the U.S. currency is facing a day of reckoning and that the only safe way to store wealth is in metals. He invests in resource companies because their stock values hinge on the prices of the metals in the ground. No market movements, not even the fall of gold's value from near $1,900 per ounce in August 2011 to below $1,250 per ounce less than two years later, will deter his investment dogma.[47] He may well be proved right in time.

Finally there is the "gambler." He is a risk taker who knows that most of his bets will fail; but he feels if he invests in enough losers, the odds are that he will win one or two. And that's all he needs. A $1,000 investment in a $.05 stock that jumps to $2 will erase the loss of forty other bad bets. The savvy gamblers diversify their investments in other industries so the losses don't mount. They chat with mining company executives, visit end users, and develop an investment thesis. They are not necessarily wiser than the other potential investors, just more informed of the risks.[48]

Most of the big-moneyed Wall Street firms shy away from investing in rare metals; the market is just too volatile, the companies too small, and the management challenges of bringing a mine into production, too daunting. Consider the various price spikes of rare earth elements indium and rhenium over the past fifteen years. Within roughly two-year spans, prices of each jumped more than tenfold before quickly losing much of their value.[49]

A Wall Street investment banker at one large firm told me that less than 3 percent of his entire mining portfolio was minor metals. With volatile pricing a hallmark of the rare metal markets, he would rather deal with larger firms like CBMM with experience and proven technologies and base metals than with a ten-man shop having aspirations and a rocky outcropping in the Arctic Tundra. Moreover, investing in these markets is guesswork: the market for these metals is small and opaque, demand for them is reliant on emerging technologies, and supply is often behind a corporate or state veil of secrecy.

"It's really just Las Vegas," Jeff Phillips, the rare metal investor, says.[50] He did well in 2010 and 2011 as he tells me when his investments soared, some by 2,000 percent, as his rare earth mines' value skyrocketed on the theoretical value of the resources they had in the ground. But then the number of rare earth projects proliferated, by some counts to more than four hundred sites. "They were never going to work," Phillips tells me. It would be problematic if even 10 percent did; they would flood the market, reduce prices, and eventually bankrupt all producers.

Because the risk-averse stay away, the seeds of our future technology are in the hands of investors who provide the millions, or tens of millions, of dollars to early stage mining companies. Phillips's money does not build the mine, but rather provides short-term—which he defines as around two years—capital to fund exploratory work. During that time, he provides management advice on how to prove the company's viability to deeper-pocketed investors, who are more willing to invest in mining projects with a higher probability of success.

Few investors seem to hold on to investments for the long term—they are looking for a good quick return. The problem in ensuring rare metal supply for our high-tech future is that the system for financing critical materials in much of the world essentially rests on a market that shuns risk and focuses on immediate gain. CBMM had long-term, deep-pocketed support, the kind rarely found in capital markets today and increasingly only available from governments. Although these resources are crucial for the future, investing successfully in any one project for the long term takes an understanding of technology, mining, and geopolitics along with patience that few investors outside governments possess.

Bringing a minor-metal mine to production takes time, especially for more complex projects such as rare earth elements, as investor Jeff Phillips well knows. The steps required to prove that a company can get a resource out of the ground and sell it are daunting. Most companies fail.

The first step in bringing a complex rare earth company to production, the one Phillips often supports, is assessing the rare earth resources in the ground—a process that can take two to five years. What is important is not just the prevalence of an element in the ground, but a concentration that is high enough for companies to be able to mine it profitably.

Once a company finds a resource, it develops a feasibility study to show investors that it understands the mineralogy of the deposit and has a framework to produce the end-stage material. The research to put this kind of study together can take more than five years and cost millions of dollars. Next, the company builds a demonstration plant, a challenge in regard to many rare metals, but especially rare earths, due to the complex chemistry needed to produce them. The company must also show that it can produce rare earth material profitably on a large scale and it seeks project partners, usually end-use customers like high-tech companies that will agree to buy the metals the company produces. At the same time, the company assesses the project's environmental impact and seeks government approvals. The total cost for these steps can be near $100 million and this is even before construction.

When I started attending rare metal conferences, I admired many executives for their attempts to pull off what seemed to be impossible. But I also felt sad for them. At conference after conference, they gave the same pitch to nearly the same group of people, most of them their peers, hoping that some new investor would throw a few dollars their way. As one representative commented to me, they knew each other's projects so well they could likely deliver each other's presentations.

But many in the industry feel CEOs need little pity. Despite the precariousness of the industry in 2013, many individual executives were making money—some, a lot of it—just for coming to these conferences year after year. They needed only a few investors to bankroll their dreams. "Their business is to sell their share price," Danny Lehrman of the metal-trading firm, Hudson Metals, tells me. "They get more money with minerals in the ground than in production."[51]

Bubar has run his company, Avalon, for nearly two decades and has yet to produce commercially salable metal. But in 2013, he made $400,000 in base salary, not including the

nearly $400,000 in stock options, according to company fil-
ings, which would be a boon if the company's stock were to
rise.[52] Because junior mining companies like Avalon don't earn
revenue as they don't yet sell anything, any dollar spent comes
directly from an investor's pockets. This is why investors might
be pleased that Avalon did not offer him a $100,000 bonus as
they did in 2011 when the stock soared on the back of high rare
earth prices. But you could argue that the real boon for CEOs
like Bubar is not the salary but being able to cash in on his 1.5
million shares of Avalon stock when Avalon's fortunes change.
He is set to do well even if the firm lets him go; his golden para-
chute would net him $1.2 million.[53]

While Bubar has to convince the market every day that he is
on track, it's a concern that CBMM's Carneiro doesn't have to
worry about. Carneiro focuses on the business of maintaining
market share. As China and India continue to urbanize and
upgrade the quality of their steel, CBMM's fortunes, along with
the niobium market, are set to grow. The demand for new
buildings along with explosion of construction over the past
decades in countries like China has led to shoddy quality
buildings, which necessitates rebuilding. Tragically, weak steel
contributed to building collapses such as in the Sichuan earth-
quake in 2008.

　　When Carneiro explains how using his company's material can strengthen steel to prevent death from earthquakes
and lead to a greener environment, you are almost cheering for
him. But he needs little support. One competitor explained it
is keenly aware that it cannot impinge too much on CBMM's
market, because CBMM has the capacity to lower prices, increase market share, and put the competition out of business.
Even with the advantage of a low cost mine, efficient opera-

tions, and the ability to dominate an even larger share of the market, there are benefits to keeping competitors around: they provide CBMM the room to match competitors' higher prices without raising the ire of its customers.

This kind of talk seems to irritate Carneiro. He tells me that he shuns those who create "myths," but he is loath to open up his company to more scrutiny to dispel them. "We don't need to be in the newspaper. We always thought that those who need to know us, know," he tells me. One of the main barriers to a reliable resource supply is opacity. To ensure a continuous stream of investment and a supply at fair prices, market information is needed. And when one private organization dominates the market as CBMM does for niobium, crucial market data are missing for the world's supplies of critical materials deterring would-be investors.

My conversation ends abruptly. Carneiro's next guests are here. I step out of the visitor center onto a new welcome mat, whiter than the one where I left my previous footprint. I hop into the car expecting a tour of the mine—often the next step after such a discussion. But we aren't authorized for a visit that day.

On our way out, I notice a series of a hundred or so white flagpoles that line the entrance road. Brazil's flag waves at the entrance, but flapping just up the hill is the flag of China followed by the flags of Japan and then Korea. I couldn't tell whether the order signified the size of country's ownership stake or the number of orders from each country. My guess was that it didn't matter. CBMM has become so successful that the line between customer and owner is now blurred.

CBMM has an incumbency advantage that no one can beat because its market is so small, its resource so rich, and its operating costs so low. But as I leave CBMM, my mind wanders to an upstart company named NioCorp, a former rare earth company that once claimed to be the largest rare earth mine until a few months earlier. It is trying to take on CBMM.[54]

NioCorp can't compete on price. The company plans to open an underground mine, which is far more expensive than CBMM's open-pit mine. In addition, NioCorp must essentially embed the cost of setting up the project in each ton of niobium it sells. To overcome the lower margins and prove itself as viable, the company markets itself as the only U.S. domestic source. The implicit assumption is that relying solely on CBMM is risky. Only time will tell whether the argument—American niobium for American Steel—will win with the market. But one thing seems certain: NioCorp will keep the dreams of some investors alive and an income stream going for management as well.

As we drive back into Araxá, I look out from above the Hotel e Termas de Araxá toward the rolling hills of the state of Minas Gerais, or "general mines" in Portuguese, named because of its rich ore deposits. Off in the distance, another mine, like CBMM, is producing a rich concentrate of minerals, but instead of processing it locally, it sends its concentrated elements off somewhere far more remote, to a frozen town where thousands of people lived but for decades it never publicly existed. And that's where the real metal-production challenges begin.

IV

Production Difficulties

Acid Washes and Talent Drains

U nlike other Soviet citizens, those living in Sillamäe, Estonia, never longed for bananas. The priv-
ileged Sillamäe residents also had the rare benefit
of owning cars, but, because of travel restrictions,
most went no farther than their jobs at Silmet, or Factory
Number 7, the town's metal refinery at the end of Central
Street. Nearly all the residents of the town, just thirty kilometers from the Russian border, were Russian-speaking, some
from republics as far away as Uzbekistan. Less than 5 percent
of the town population was native. The town lacked a local
identity, but it had a purpose. The scientists who worked at the
facility were engaged in top-secret metallurgical processes,
refining uranium, rare earths, niobium, and tantalum.[1]

Joseph Stalin built the city in the late 1940s to be the pride
of the Baltic region. When I visited in 2013, it still retained a veneer of its grandeur, but it showed its seventy years of wear. The
city's architecture was bold and commanding: tree-lined bou-
levards; two- and four-story neoclassical buildings with stately,

well-worn facades; and plenty of green space, including a park with a cascading staircase adorned with urns and gargoyles.[2]

Sillamäe has stunning vistas overlooking the Gulf of Finland. But for decades, visiting it was impossible. Once construction of the town was complete—a feat that took up to eighteen thousand people roughly five years—Stalin erased it from official maps. Supplied with its own dedicated water and electricity systems, the town and its metal-processing facility became a state secret. Stalin chose to build the processing facility in Sillamäe because geologists had reported that the town was rich in uranium; some of it was used to make the country's first atomic bomb. But, despite the Soviets' meticulous planning, Factory Number 7, now known as Silmet, was built in the wrong spot. Sillamäe ran out of uranium not long after the mining started. Rumor in town had it that scientists had overestimated the resource buried in the region's vast oil shale reserves. One wonders whether some unlucky souls spent years in Siberia contemplating this misjudgment.[3]

After the town's ore ran out, Silmet started importing it. This imported uranium ore contained contaminants, other metals that complicated the refining process. Instead of discarding these impurities, notably niobium and tantalum, the Soviets set up two more processing lines to extract them. When alloyed with other metals, niobium and tantalum strengthen them to withstand extreme temperatures, which prevents them from cracking in places like plane turbines. At the time that Silmet built its new lines, tantalum was also emerging as a crucial component in capacitors for various electronic and defense applications because of its unique ability to store an electric charge.[4]

Unfortunately for the workers at Silmet, rare metals don't neatly separate out of the soil like pebbles from sand. Mining companies need to find ways to turn sand and crushed rocks into metals because few appear in the ground in their elemental state. (Gold is a notable exception.) Enter beneficiation, concentration, cracking, extracting, refining, and metallurgy, the process of turning ore into metal through a combination of steps that include floating, roasting, and diluting rocks in acid, as we saw briefly at CBMM. Many rare metals are so technically challenging for chemists to produce that it is better to think of them as chemical creations rather than geological minerals.

Processing rare metals is all about trade-offs—which metals to process and which to discard. Because by-product metals are produced in small amounts and need their own processing equipment, they are often discarded. But the only way that many rare metals are produced is as by-products because their concentration in the ground is so low that miners cannot profitably mine them directly.

Just to develop the metallurgical processes to refine the minerals from a new mine, can cost up to hundreds of millions of dollars. Even spending such high sums does not guarantee success. Each step has to be perfected through sampling and adjusting before work on the next stage can begin. Refining rare metals requires a balance of time, temperature, and liquids. It's unforgiving. And there are no assurances that companies can keep the processing costs low enough to make new mine sites profitable. Some companies even have the rare metal–rich minerals, but cannot find the right heat and liquid combinations to coax the rare metals out. Processing is the hidden side of mining. Having the ore is just the beginning.

There are three main steps to refining rare earth elements and indeed many rare metals: increase the concentration of the elements from the ore; extract the elements from the concentrated mix; and separate individual rare earth metals. No step is inexpensive or straightforward; add in too much sulfuric acid and you dissolve too many minerals, making it harder to extract the rare earth elements. And the risks are not only financial. "If you make a mistake, you're dead," says Richard Hammen, chief executive officer (CEO) at IntelliMet, a rare earth technology company. In fact, the processing of some materials can itself be downright deadly. In Andover, Massachusetts, a 55-ton furnace that applies heat and pressure to metals blew up, sending shrapnel pieces of nearly 3 tons more than a quarter mile away.[5]

You can't just open up a cookbook for metals and make neodymium or dysprosium. "There is no school where you can learn solvent extraction," says Alain Leveque, the former global research and development (R&D) director for rare earth elements at the processing firm Rhodia (now Solvay), referring to the chemical process for rare earth separation. "You have to develop your own knowledge or develop an agreement with someone who already has it." Leveque should know; Rhodia devised many of the processing techniques that are currently used. The definitive guide on rare earth processing is titled the *Extractive Metallurgy of Rare Earths* and is filled with flow charts and chemical formulas but cautions, "the processes actually practiced in the industry are well-kept secrets." And those in the know are not always willing to share their knowledge.[6]

Rare earths come in more than 160 different minerals but metallurgists have perfected extraction of the elements from just a handful. What's more, because every ore deposit is unique, Hammen tells me, the combination and amount of chemicals used to produce concentrated rare earth solutions from each deposit are equally unique, although the process has some standard steps.[7]

The first step is to dissolve the crushed minerals, the rocks that contain the elements, into a mix of acids, such as hydrochloric acid, to make a concentrated mix of rare earth elements. Scientists like Hammen need to find the unique balance of acids that dissolves rare earth minerals the way salt dissolves in a glass of water.

Sometimes an insufficient quantity of the rare earths is dissolved, meaning a high percentage is wasted; other times the rare earth mixture turns into a viscous blob, making the extraction of sufficient elements from it difficult, if not impossible. Because most of the minerals going into this mixture contain less than 10 percent rare earth elements, the liquid solution is not transparent. In many cases, the mixture contains ten times more iron than rare earths, which turns the solution into a rich brackish brew that also happens to be radioactive because of the naturally occurring elements in the mix. Setting up the process to get just this far can take years of adjustments.[8]

The next step is to filter out the contaminants and extract the elements from the concentrated rare earth soluble mix with a centrifuge or a filter. The goal is not so much to remove rare earths from the mix as to remove everything else from the mix. As the contaminants are taken out, the liquid slowly becomes clearer. Because some of the contaminants, like the rare earths themselves, have been absorbed into the acidic brew, chemists will add other acids to change the mix's acidity levels so as to turn the contaminants back into a solid. The contaminants are then filtered out, leaving a rare earth and acid mix.

The acid is then evaporated, leaving a gray powdery concentration of mostly rare earth elements. The final step is to process the concentrate into individual rare earth element powders. This last step is what Silmet, the processing facility in Sillamäe, did in Soviet times and continues to do today.

About five hundred miles away from Sillamäe and roughly 230 years ago, while he was walking around a feldspar quarry in the town of Ytterby in 1787, Swedish army lieutenant Carl Axel Arrhenius found a curious black, dense rock. Its appearance seemed out of place in the mine. As a chemist he took it upon himself to analyze its composition and soon realized he had found a "new" mineral, which he called ytterbite (now gadolinite). But it wasn't until 1789 that the Finnish chemist Johan Gadolin extracted the first rare earth element from the rock, yttrium. But finding successive elements in the rock was challenging. It took about ten years for Jöns J. Berzelius, Arrhenius's professor, to discover another rare earth element and another fifty years for the next to be uncovered. In fact, these metals confounded scientists to such an extent that for most of the 1800s physicists believed that didymium was one element until Carl Auer von Welsbach discovered in the 1880s that they were actually two. Chemists needed a total of 150 years to isolate all 17 rare elements, the last being promethium in 1947.[9]

The challenge scientists face in isolating all the rare earth elements is that they have similar atomic configurations, which makes identifying and separating the element's tight chemical bonds they form with one another difficult. Scientists have had to isolate each successive element through slow chemical and heating processes to strip off the known elements and isolate the targeted ones. Today, the workers at Silmet go through similar steps to separate the rare earth elements they produce.[10]

Historically, Silmet has separated 3 percent of the world's rare earth output, which adds up to several thousand tons annually. But that 3 percent was the only substantial amount of rare earths produced outside China for much of the early 2000s. Despite its unique market position, Silmet faced economic difficulties during the same time period when China began producing rare earths at increasingly lower prices.

Silmet's campus of brick buildings has changed little since Soviet times. Pipes meander overhead from one building to the next as if in an M. C. Escher painting. The rusted veneer on many of the pipes and the peeling insulation wrap on others make me think they've lived past their usefulness, but my guide, David O'Brock, Silmet's CEO, assures me they all work because he would melt them down for scrap if they didn't.

One of the first stops on his tour is an empty oversized control room lined with square gray control panels replete with lights that no longer light and switches that no longer flip. A desk sits in the center, equally boxy, on which more toggles and buttons reside in front of and on either side of where an operator would sit. The company was going to take apart the room but for its nostalgic obsolescence. It reminds O'Brock too much of a set from a Sean Connery-era James Bond or Austin Powers movie and he wants to keep it to remind the company of its roots.

For decades, the socialist system rewarded Silmet for production, not for sales, because the company's wares had an immediate buyer: the state. The company had produced its metals, but had no idea what they were used for. O'Brock had to find out what his materials did and who might buy them.

O'Brock is an unlikely head of a former Soviet secret facility. He grew up on a farm in Ohio and, on a lark, came to Estonia in 1994 to study. He now has a wife and two children and speaks flawless Russian and Estonian. The company hired him, he says, to put Silmet on sound financial footing.

O'Brock quickly found new customers when he came on-board in 1999 and then eventually brokered the sale of Silmet to the U.S.-based Molycorp in 2011. When we met, I quickly understood why customers took to him: he has an easy way about him and a boyish enthusiasm when he speaks about his company.

After visiting the antiquated control room, O'Brock takes me into one of the largest red brick buildings in the Silmet complex, standing at around seven stories. Sitting in front of me is a maze of large steel pipes that rises and falls above rows and rows of stainless steel rectangular containers the size of oversized garbage cans sitting several feet off the cement floor on metal scaffolding. Each one of the three hundred containers has a red number emblazoned on it and protruding thick welded-metal tubes that connect one box with the next.

O'Brock explains to me that inside each one of these boxes, known as extractors, is a mix of rare earth elements and acid. Because rare earth elements, and indeed all elements, have slightly different weights, the rare earths will separate at different levels inside the extractors. The extractors then filter the liquids at different depths to collect as much of the desired element as possible. Since an individual extractor is not particularly efficient, Silmet must process the mix three hundred times just to separate out liquid from the mix of the first two rare earth elements they produce, cerium and lanthanum, the most plentiful and least expensive rare earth elements.

From the original batch, the processing continues yet another three hundred times to separate out the other two elements that Silmet sells, praseodymium and neodymium. It may seem evident but Silmet can only produce the assortment of elements that reside in the concentrated minerals it receives. But since demand for these elements does not neatly equate to their distribution in the ground, there is often a glut of some elements and a shortage of others. What adds to the supply-and-demand mismatch is that processors like Silmet cannot just separate out the high-priced elements because there is a prescribed order to removing the elements.

As we will see, Molycorp has a problem similar to the problems of a number of other rare earth producers outside China; roughly 80 percent of the company's rare earth reserves are in low-value cerium and lanthanum, which must be produced to access more valuable elements. Furthermore, Molycorp's Silmet refinery only produces one set of rare earths commonly known as "light" rare earths, referring to their lower atomic numbers, but it might as well refer to the amount of cash they bring in. The other group of rare earth elements, the "heavies," is more valuable and produced in very low quantities, nearly all from southern China due to the region's favorable geology.[11]

Producing the high-grade materials increasingly demanded by industry for high-tech, green, and defense applications is an expensive and challenging task. In fact, Sillamäe is unable to produce the highest grade rare earth materials. Their process for refinement, although good for producing bulk material, does not work for separating and producing the high-purity, light rare earths that the market demands. That technology lies in China.[12]

To process the rare earth element powders that Silmet produces takes about two weeks, but many rare metals take far longer. Producing lithium—a soft metal found dissolved in saltwater deposits, and best known by consumers for its use in batteries—can take twelve to twenty-four months.[13]

Some rare metals are not even produced directly by processing minerals buried in the earth; germanium comes from the residue of coal ash. And scientists are working on novel ways to extract metals from improbable sources where concentrations of metals are very low, such as from wastewater and even from roadside plants that absorb palladium from car exhaust.[14]

When geologists say that "ore grade is king," they mean that the richer the metal content of a deposit, the more profitable the mine will be. That is generally true for gold or copper because a richer ore will require less grinding and less processing, which are the costliest parts of production. But a focus on the richness of the ore alone is unhelpful in understanding the economics of a rare metal deposit and the importance of processing. To understand those issues, one must look at the efficiencies and the costs of production.

Mining companies lose material in every step of refining, from taking the mineral out of the ground to turning it into metal. Minerals are lost even before mining begins because engineers cannot design a cost-efficient mine plan that allows digging equipment to extract from every nook of mineral deposits. Minerals are also lost in blasting and when they fall from the truck in transport. The next steps—crushing the ore and putting it in the acid baths that help to turn mineral ore into a concentrate—sacrifice even more precious material, as does the separation of rare metals from other metals in the mix.

It is instructive to look at how this loss plays out. Frontier Rare Earth company plans to dig up 20 million tons at a rare earth mine in South Africa called Zandkopsdrift, a dusty barren hump sitting on a rustic-orange rolling plain. But the company estimates that it will recover only 620,000 tons of rare earth material. Based on the ratio of material to ore, the deposit is a 3 percent rare earth grade. Compared to some, this is a low-grade deposit, although it's full of high-priced metals.

One can't assume that the company will eventually sell all 620,000 tons. Frontier will further lose about 10 percent of the rare earths in mining, 24 percent in concentrating the ore, and then another 10–12 percent while separating and refining the mineral into rare metal powders. In the end, the processing will yield only 375,000 tons of rare earth powders over the life of the mine—just 60 percent of the total rare earth elements in the ground at the mining site.[15]

Because of these challenges, some of the lowest grade rare earth deposits, for example, are some of the most profitable. In the southern Chinese province of Jiangxi, the rare earth ore grade is extremely low—less than 0.2 percent of the ore is rare earth elements compared to a grade of 8.2 percent at Molycorp's Mountain Pass mine, the only producing rare earth mine in the United States.[16] What makes the rare earths around Jiangxi profitable is that the soil is brittle clay, weathered by millions of years of exposure to a hot, wet climate. The weather has worn down the mineral deposits so they are loosely bonded to the surrounding clay. The result is an extremely fine-grain mineral deposit that is easy to dig up. More importantly, the ore is inexpensive to process. The locals just need to dig up soil from the hillside, dump it in acid, and bake it at high temperatures to produce rare earth concentrate. And because environmental laws have been historically lax, locals have faced little need to invest in environmental protection. Processing rare earths is so easy that farmers in southern China, after tending the fields, can make a good second income.

In northern China near Mongolia sits Baotou's Bayan Obo mine, the world's largest producing rare earth mine. Only 6 percent of its ore contains rare earth. Despite this relatively low grade, it is the world's lowest-cost rare earth producer. Most of Bayan Obo's rare earths are locked in the same type of minerals as those in Molycorp's mine. But the Chinese produce rare earths more cheaply because Bayan Obo is not solely a rare earth mine. It is an iron-ore mine with a very profitable rare earth side business. Since Bayan Obo is already digging and processing the ore to remove iron, processing rare earths there involves half the cost of doing so elsewhere.[17]

In fact, rare earths, like most minor metals, are often by-products of other mined base metals. For example, processing aluminum and zinc yields gallium; nickel and copper deposits produce cobalt; and zinc yields indium. There are roughly a dozen trace metals in copper alone. Having by-products in a deposit may appear to be a boon to a company's bottom line; however, many companies don't celebrate such finds. Rather, these rare metals are a nuisance, impurities that must be removed, which increases the metallurgical costs of processing and leads to wasting of elements that would be considered valuable in other contexts.[18]

In 2011, refiners processed only 585 tons of tellurium, an element named for the Roman earth goddess, Tellus. Part of the reason for the limited quantities is because of its availability: it's four times less common than gold in the earth's crust. Nearly all current supplies come from copper wastes, which raises an economic problem. Because by-product metals are not directly mined, they don't respond neatly to the laws of supply and demand. Unless companies have easy-to-access supplies of tellurium-laden copper waste around, a higher price for tellurium doesn't often provide enough incentive for copper miners to produce more of it. It's not valuable enough.[19]

A joint U.S. National Renewable Energy Laboratory/ Colorado School of Mines study found that the value of the copper ore at a mine is thousands of times greater than that of tellurium. The authors note that tellurium supply, especially over the short and medium term, does not respond to the demand for tellurium but rather to the demand for copper. Over a ten-year period beginning in the early 2000s, tellurium prices increased tenfold, but copper industry production of tellurium stagnated.[20]

The rare metal is recovered for processing from high-grade copper-rich ores only—an increasingly limited source because copper-ore grades are falling. Changing a low-grade copper refining system to maximize tellurium recovery would mean building an entire new processing system that would sacrifice copper output. This isn't a decision that a mining company would even consider; there is little money in tellurium because the market is small—in 2012, the hundreds of tons of tellurium produced had a market value of around $100 million, whereas copper's 17 million ton market was worth $136 billion.[21]

In addition, recovery from secondary sources is inefficient. For example, processing companies recover only 20 percent of indium from the host metal, zinc, because at each stage of the zinc-refining process indium is lost because the economic incentive to maximize indium is minimal. The authors of the U.S. National Renewable Energy Laboratory found that because of the peculiar nature of by-product production, supply for metals like indium and tellurium react in a stepwise fashion with price increases, not incrementally as is the case with commodities such as oil or coal. Prices need to cross certain thresholds to encourage a new, more expensive method of production.[22]

For example, the authors estimated that if the price of indium was about $300 per kilogram, current refiners could produce between 1,800 and 2,900 tons at a profit, which means a higher price may induce some companies to produce more with a small price increase. However, the price of indium would have to jump to about $600 per kilogram for companies to produce 3,000 tons because companies would need to make more expensive investments in processing. Moreover, the authors concluded that additional production would not hit the market for at least five years, meaning that if, for example, the demand for solar panels or flat screens using indium spiked, supplies would be stretched thin.[23]

As we walk around the decaying Soviet-era factory grounds on this early January afternoon under a setting, weak winter sun, O'Brock tells me how managing three different rare metal production lines—rare earth powders tantalum, and niobium—has helped Silmet to survive in a post-Soviet world. The diversity of its products ensures that one of the metals is always at a high enough price point to subsidize the production of the others.

The stern faces around the Estonian plant are far different from the smiles I encountered in Araxá, the niobium mining town in Brazil. Despite the regions' geographic and cultural differences, snow-dusted Sillamäe and the verdant hills of Brazil are one stop—a cargo-ship ride—from one another on the niobium supply line. The crushed ore from which O'Brock produces his metals comes from the same area of Brazil as the niobium giant CBMM's mine.

Because CBMM meets 85 percent of the world's niobium needs due to its low costs, it has the ability to squash niobium producers like Silmet. But Silmet has found a small profitable niche. The Estonian refiner sells small amounts of specialized high-grade niobium metal to universities and small manufacturers, rather than to giant multinational companies. Silmet's metal finds its way into magnetic resonance imaging machines, televisions, and even the electromagnets that steer streams of protons around CERN's Large Hadron Collider, the world's highest-powered particle accelerator, which is located in Switzerland.[24]

O'Brock doesn't like to show the initial processing of niobium from the Brazilian supplies. He is proud of the safety of his facility but says the building is still dangerous, since the processing produces a lot of volatile dust. "Every year one factory in China pretty much blows up," he tells me. He adds that the factory of one of his rivals in Brazil caught fire in 2009 for getting the process wrong.[25]

Instead, O'Brock brings me into a facility where shiny two-meter-long metal blocks lie on a steel rack. Silmet mixes 94 percent niobium powder with 6 percent aluminum to help convert the niobium to metal. Whereas rare earths rely on mixes of acid to process the metal, high-grade niobium needs heat and lots of it.

Workers slide the niobium-aluminum ingots into a long blue tube of a furnace that shoots a white electron beam at the metal, heating it to 2,300 degrees Celsius (4,172 degrees Fahrenheit). That temperature vaporizes nearly everything except the niobium, which turns into a liquid and collects at the bottom furnace.

Producing high-purity metal is expensive due to the often long process of alternating a mix of high temperature and acids. Not only does the equipment cost tens of millions of dollars, operating at such high temperature uses vast amounts of electricity. And despite the best efforts of O'Brock's employees, impurities are everywhere—in the acid, in the furnaces, even in the air. Unwanted elements such as oxygen are always creeping into the final product, undermining efforts to produce the high-quality, high-priced material that Silmet's customers demand.[26]

Throughout our tour, I am struck by the age of some of Silmet's equipment. The niobium furnace, for example, also has a Soviet-era control panel that resembles an Atari video-game system. Despite the antiquated tools, O'Brock boasts that his metallurgists produce higher-grade material more consistently than do companies with newer equipment.

While the distance from Sillamäe to Jiangxi Province in southern China, the heart of heavy rare earth country, is 4,500 miles, few points on the planet could feel farther apart. As I approached the entrance of an oversized concrete hut, set on the slight downward slope of a hill, a waft of acrid warm air greeted me, overpowering the already oppressive 90 degree summer, humid heat. Upon entering, I looked up and saw a rectangular patchwork of thick tree branches supporting a tin roof. If the walls of this building were mud, I felt it could be a church in a poor, rural African village. Along the length of the building sat ten small furnaces with a bubbling orange glow emanating from a hole on top of each. Behind every furnace dangling from a hook was a laminated paper sign with a few letters in red writing: HO or DY (holmium and dysprosium) to indicate the specific metal boiling in each.

At this rare earth smelter nestled in the hills not far from where the material is mined, almost all the workers sported light silver-beige pants and nearly matching jackets with front zippers, which some men actually used. Most of the workers wore facemasks, not the ones that actually protect from nasty toxins, but the kind dental hygienists wear to ensure that they don't get splashed. Those who weren't wearing masks had cigarettes dangling from their lips. Because their hands were busy—banging a hammer to brand the metal or pouring powder from a metal soup ladle into the coffee-cup-size cauldron that would be placed in the furnace—they were prevented from really smoking. They paid little heed to the presence of foreign visitors.

The building had little ventilation to remove the heat or the fumes, save a slight natural breeze. Above the furnaces along the top of the wall was a long ventilation hood hung ten feet above furnaces. To me, it seemed a perfunctory tool to guide away the fumes rather than a serious attempt to keep the air clean. If Silmet was considered low-tech, this was old-tech.

But despite a seeming lack of technical sophistication, this Jiangxi smelter and others nearby are the material suppliers for the world's highest technological applications. The smartphone in my pocket had its roots in one of the low-tech cauldrons I was staring at. As I tried to make sense of my gadgets' DNA, my throat and nose started to burn. My Japanese colleague told me it was because of fluorine gas. Sometimes people are affected by it, he said. It is not healthy. I later found out that exposure can severely damage the eyes, skin, and the entire respiratory system. In some cases, it can be fatal.[27]

We promptly left the building and went to some of the other smelting buildings nearby, some with larger capacity furnaces and vacant concrete floor space next to them. The director told me he was waiting to add more processing equipment to increase the company's capacity. These buildings looked just as antiquated as the others I had seen, but he asked me not to take pictures.

A couple miles away from the metal refiner sat a rare earth separation facility, similar to what I'd seen in Sillamäe, but with slightly newer equipment housed in a tin-roofed warehouse. What was most striking about the facility was not its equipment, but that it is was lying fallow. China had so much extra refining capacity because companies had overbuilt, and at the same time the government was cracking down on illegal mining, leading to less rare earth minerals to process. Overcapacity in China, not shortages, was the sword of Damocles over the rare earth industry in 2013.

Start-up rare earth mining companies aiming to compete against China have to realize that even if they could raise hundreds of millions of dollars to start up a new facility, at any moment producers in China can open shuttered processing lines and lower prices around the world, putting those new investments in peril. The new mining companies can have all the advanced technology money can buy, but if they can't compete with Chinese operating costs, it matters little. China was cutting down on illegal metal production for a myriad of reasons—from ensuring that companies adhere to environmental and worker safety regulations, to reducing supply in order to increase prices for rare earths. As I walked around the processing site, I saw a sheared-off hill with a brown open expanse in front of it just beyond the boundaries of the current compound. The director told me he had plans to double his processing capabilities to become more profitable and needed the extra space.

Despite the importance of turning minerals into metal, regrettably, there are fewer and fewer experienced practicing geologists and metallurgists, especially in the West. The average age of the mining workforce in the United States is close to fifty; most will retire in the next ten to fifteen years. Likewise, in Canada, industry sources estimate that nearly 60,000 workers may retire by 2020, but the sector may need 100,000 new workers to meet demand. It took Avalon Rare Metal two years to find a vice president of metallurgy, and during the search, the company relied on two metallurgists, a septuagenarian and an octogenarian, to support their work.[28]

"All kids are geologists," Mia Boiridy told me. "At some point, they all bring rocks home." As the director of the Dynamic Earth Program at the educational organization Science North in Canada, Boiridy educates children about the wonders of the minerals around us. Somewhere along the way to adulthood, most children lose their interest in rocks. Life speeds up, and people spend less time looking down and more time looking ahead. "There are limited [numbers of] people who practice this art," laments Corby Anderson, professor of metallurgy at the Colorado School of Mines, himself a third-generation mining hand. This has profound implications.[29]

The lack of a knowledgeable workforce, even in countries with strong mining histories like Australia and Canada, will mean a slow development of local mines in Western countries and a growing reliance on mines developed in China, Russia, or Brazil, by Chinese, Russian, or Brazilian engineers. It also means that new young Western workers will have increasingly fewer mentors, crucial to these young scientists in developing their skills and providing them critical information.[30]

Steve Yue, chairman of the Department of Mining and Materials Engineering at McGill University, notes that it is increasingly difficult to lure young students into metallurgical studies and very hard to find new professors.[31] This difficulty recruiting students combined with the fact that engineering and metallurgy are as much experience as they are science, the bottleneck to producing rare earth resources, especially in the West, may turn out to be not geologic or economic, but human. Corby Anderson believes that the lack of trained mining engineers and metallurgists could affect our ability to produce the materials we need in the years ahead.

But the Anderson clan is helping. Anderson's son, Caelen, works just a few doors down from him. Caelen grew up shoveling dirt at a copper mine and quickly learned he would rather design mines than dig in them. He is a budding hydrometallurgist, a chemist who uses liquids as opposed to heat to separate metals. He tells me he is one of just a handful of future PhD hydrometallurgists that Western universities will produce over the next five years, meaning he will face little competition for jobs.

The problem is that the unheralded art of metallurgy gets short shrift, even on the well-manicured grounds of the Colorado School of Mines, where mining is literally in the name of the school. The campus's showcase building—for petroleum engineering—has nothing to do with rocks. According to the campus donation board, ConocoPhillips, Hess, and Schlumberger donated at least a million dollars each to ensure that this state-of-the-art building was loaded with flat screens and conference rooms filled with small Asian vases visible behind protective glass. The companies' names appear throughout the building: Halliburton and Hess sponsor adjoining classrooms; Marathon Oil funded a center of excellence for reservoir studies. Even the LED lighting underneath the railings, leading up from the three-story, glass-enclosed entrance, show that the school spared little expense.

Across the campus in a nondescript beige brick building is the home of the metallurgy department where Caelen and his father work. In fact, it is a jumble of buildings. The office is dark, even at midday, but functional and definitely a few steps up in quality from the aging environmental studies annex where white fluorescent lights still hum loudly and the over-painted paned windows have yet to be replaced by energy efficient ones.

But Caelen is fortunate: he is at the only mining school in the country that has a rare earth specialty.[32]

The challenge to producing the rare metals that the world needs is that the Anderson progeny and their colleagues will require more support. The world needs a robust set of metallurgists to unlock the secrets of extracting elements from a myriad of minerals, not just the ones we are currently mining.

Silmet does its job admirably, but I would argue that it is time for an upgrade—not just newer equipment, but newer ideas. Innovations in processing, designed by educated and skilled workers, can bring far more materials to the market, more cheaply, than can mining companies just searching for new deposits. New sources of metal, whether these are new mineral deposits, old mining waste, or even wastewater, could help reduce the world's reliance on the limited set of deposits we have discovered.

But policymakers, academic institutions, and even investors often overlook the processing challenges, as if the companies need only the minerals themselves to be productive. Until metallurgists find more efficient ways to process rare metals,

Silmet will continue its metallurgical tradition that extends back to the dawn of the Soviet Union. Fortunately for O'Brock, he knows he has buyers for his product. Still, because of the web of traders that are the hidden link between the processors and the product, he may never know where his rare metals go.

V

Trading Networks
Smugglers and Supply Hiccups

Have you met Super Mario?" the head of research at a Japanese trading house asked me. "You must meet him." He told me that Super Mario, the famous Italian plumber of the Nintendo video game, was also a nickname for the main supplier of rare earth metals to Japanese companies, Shigeo Nakamura. He is one of the unsung heroes of Japanese industry; the one responsible for bringing the country's high-tech industries rare earth elements to polish glass for their camera industry and phosphors for their flat screens. He is the critical connection between processors like Silmet and Japan's high-tech manufacturers. Nakamura is also part of a small, tight-knit fraternity (traders are nearly all men), where people earn hefty sums just trading metals like indium or ferro niobium.

Most of the large Japanese trading houses such as Mitsui, Marubeni, and Sumitomo shun minor metals. It's a nuisance for a trading house that provides millions of tons of base metals to supply a few tons of neodymium, a rare earth metal, or hundreds of kilograms of tantalum, and it's far less profitable. The reluctance to trade in these materials left space for Super Mario/Shigeo Nakamura to build his company, Advanced Materials Japan, into a specialty metal-trading house that has become the country's leading rare earth element supplier. He is part of a web of small companies of specialty traders that help form the backbone of the global rare metal market.

Nakamura gets his nickname from his salt-and-pepper, bushy mustache, similar to the Nintendo character's, and his jolly demeanor. Nakamura is also a character in his own right; a set of comic books romanticizes his exploits in bringing resources back to Japan—a captain of Japanese industry in a country that has made it a national priority for domestic traders to import 50 percent of the rare metals the country uses.[1]

Advanced Materials Japan sources its main products from the rolling hills of southern China or the steppes of Inner Mongolia. But some traders get metals anywhere they can; for some that means sourcing material in Vietnam that may have been smuggled out of China. Many clients don't really want to know where the metals are from; they just want a consistent supply of inexpensive resources that meet their specifications. This laissez-faire attitude provides leeway to traders to access metals in various forms of legality.

The network to get rare metals from the mine to your laptop travels through a murky network of traders, processors, and component manufacturers. Traders are the middlemen who do more than buy and sell rare metals: they help to regulate information and are the hidden link that helps in navigating the network between metals plants like Silmet and the components in our laptops.

Although personal relationships, reliability, and quality are valued, price ultimately reigns. The rare metal market's opacity combined with the limited number of mines and processors makes for very tenuous resource supply lines and volatile prices. No one really knows the true size of these markets. Even the U.S. Geological Survey, which tracks the market size of every metal, won't hazard a guess on metals markets of some minor metals. This lack of transparency adds to the volatility of markets and allows rumors and hearsay to fill the void of data and analysis.

What's more, most critical materials must meet such exacting standards that they are more akin to the products they are in, like microprocessors, than they are to commodities like copper or iron. Unlike those commodities, which are easily traded in standard amounts on exchanges, rare metals are traded in backroom deals, often in small quantities and tailored grades for specific end uses. In certain ways, the minor metal trading system is so antiquated that they trade in the way that commodities did before exchanges existed.

In countries with weak legal enforcement, smuggling can quickly take hold, in some cases fueling conflict. If there's one weak link in the supply chain, the market will seize up—prices will either skyrocket or plummet. The ability of minor metal traders to navigate the murkiness of the marketplace—while at the same time contributing to it—makes traders like Noah Lehrman indispensable.

Noah Lehrman speaks about economic reforms in China with the fluency and conviction of a Wall Street analyst. But with a scraggly orange-brown speckled beard and hair that goes well past his ears, he appears every bit the existential singer his promoters boast. He is likely the only person in history to both perform at the Jewish Grateful DeadFest and advise the U.S. Congress on resource security. Lehrman is in the family metal business and part of the same network of small metal trading shops to which Nakamura belongs, where good interpersonal relationships equate to a stable business.

Lehrman is also the U.S. representative for the Minor Metals Trade Association, the main organization for rare metal traders. Forty years ago, at the dawn of the Rare Metal Age, around the time when calculators were becoming the first handheld electronic computers, a group of young metal traders started selling cast-off metals from refiners. They were selling to scientists and electronic companies, just trying to make a few extra bucks for their producers. They banded together to form the Minor Metals Trade Association. Maria Cox, the association's current executive director, tells me that the association likely started out, as did other similar specialist groups, as a chance to share a drink with friends, and it's now the main body representing traders and has 140 member companies globally.

Lehrman's office has a homey feel despite being on the eighteenth floor of a building that is just a short distance from New York's Grand Central Station. Yankee paraphernalia covers the walls and chunks of metal and a handful of die-cast airplanes in take-off position, including the late British Airways Concorde whose last flight was over a decade ago (2003), sit on a wooden bookcase revealing not only the company's customers but the length of time it has been in the business. The office administrator has been here for twenty years.

"During the height of the Cold War, my dad had people from Red China and the Soviet Union over for dinner," Lehrman tells me. His father, Danny, born in Bolivia to a German-Hungarian father and French mother, helped open global minor metal markets when he was with his uncle's business. In the late 1970s, he ventured off to China and the Soviet Union and became the first person to import chromium from these countries. After a few successful years with his uncle's firm, Danny founded his own shop, Hudson Metals.

Like Nakamura in Japan, the Lehrmans are the connection between the Chinese tungsten and Russian titanium suppliers and the companies that make airplane engines and medical devices. With his focus on Asia, most of Noah's work entails calls made after midnight in Mandarin to Chinese traders. But you get the sense that even if his work didn't demand such late night calls, he might just be awake then anyway.

Every few months Noah travels deep into China, to remote cities that few foreigners see. He has developed connections that go beyond business. One Chinese family—knowing that Noah keeps kosher, but not really knowing what it means—has a special set of pots they use to cook his food when he visits. Maintaining such relationships is crucial for a business that relies on its reputation for quality metals.

When I questioned Noah about the apparent lack of high-tech processing equipment I saw in Jiangxi, he assured me (based on the several months a year he spends in China) that the equipment is technologically appropriate: "We're making things hot and smashing them." You don't need the fanciest equipment to do that, he tells me. Like David O'Brock, chief executive officer of Silmet, Lehrman puts more faith in the craftsmen who churn out these metals than in their having the latest tools. Both men believe that metallurgists understand the process in a way that highly mechanized equipment cannot. Being able to consistently deliver high-quality metals is important to the Lehrmans.

"We are 100 percent specification-driven," Danny Lehrman tells me. Unlike traders in other commodities, such as oil, who sit behind computer screens anonymously trading barrels of homogeneous oil on exchanges, Danny has to meet the increasingly demanding specifications of his customers. He cannot simply trade metal of a standard grade; he has to meet the quality his customers need. One substandard batch of metal ruins the relationship. It is not a risk he can take. Maintaining long-term supplier relationships and doing spot checks lessens the risk that a bad batch will reach the clients. As the Russian axiom, attributed to president Ronald Reagan, goes, "Trust but verify."[2]

The market is tough, especially in Asia. Many traders tell me Chinese firms prove to be challenging partners—contracts are looser and terms can change, especially when beginning new trading relationships. As Michael Rapaport, a rare metal trader who once worked with the Lehrmans, told me, "The Chinese don't abide by the same rules especially if the market moves out in your favor." They will back out of the contract. Rapaport avoids long-term agreements and takes delivery immediately. It's a common sentiment in the industry.[3]

Unlike base metals and commodities like oil that trade on exchanges with clearly listed prices and amounts, there are no universally accepted benchmark prices for most rare metals. The market for many minor metals, like niobium, are dominated by a few producers who keep their data proprietary, which makes finding the true price challenging. With trading taking place behind the scenes, the Lehrmans are the best source for the latest price and production trends for the materials they sell.[4]

Metal-Pages has one of the most robust price listings in the industry on its Web site, but it relies on buyers and sellers— hardly disinterested sources—to report their prices. If that process sounds as if it is ripe for manipulation, it is. If global banks were cavalier in illegally manipulating gold prices and the LIBOR—the interest rate that banks use to lend to one another, which underpins hundreds of trillions of dollars in loans and other securities—for their own gain, it would be hard to imagine that traders sitting in London or Ganzhou would not be tempted to misreport a price in their own favor.[5]

Despite the challenge of backroom deal making, a lot of the traders are quite happy for the market to be opaque, a Minor Metals Trade Association (MMTA) member told me. It offers them a profitable position. Unlike others in the market who just buy or sell, traders know the market's true metal prices as well as transactional information. And as they have personal relationships with both the seller and buyer, it makes it easier to buy low, hold material, and then sell it at higher prices. Such advantages were on the minds of MMTA members when they balked at a proposal to work with the London Metal Exchange in 2009 to establish an "online price discovery system"—60 percent voted against the proposal.[6]

There was a reluctance to change the nature of business, an MMTA member commented to me. Open prices would undermine competitive advantage. An eventual exchange could also allow manufacturers to reduce their risk through hedging, essentially buying future amounts of a commodity at a fixed price, called "futures." But at the same time an exchange could undermine the personal relationship an end user has with its trader—who ensures reliable delivery—which may ultimately be more important to a customer's long-term resource security than hedging and saving 5 percent. The Lehrmans provide this security of supply along with market advice to their customers. It is part of the reason they consider themselves merchants, not traders—that's too pejorative a term.

Danny Lehrman now tells his clients that he sees the market changing. Although minor metals trade under various contracts, about three-quarters trade under some long-term contract, according to Nigel Tunna, head of Metal-Pages news service and a former antimony trader. Danny sees businesses increasingly preferring these longer-term deals—five years and longer.[7] "They now want to lock in prices," he says. They are tired of the supply chain concerns and will sacrifice price for reliability.

With the increasing price trend for many minor metals, some outside investors are trying to get a piece of the market. The challenge for them lies not in buying rare metals because they can easily buy online at the Chinese web site alibaba.com.[8] The hard part is to sell them—this requires relationships. Boeing is not going to buy titanium alloys from a random investor with a few kilograms or pounds. Not just individual investors are trying to get a piece of the action, but banks, such as Brown Brothers Harriman, have become MMTA members and are trying to find ways to work with traders. But no group has been as successful as Chinese exchanges in making profits while changing the way minor metals are traded. But their business model raises concern about stability of the market.

These budding rare metal exchanges geared to Chinese investors who are struggling to find profitable domestic investment opportunities are the new rage. Since wealthy Chinese have limited investment options because the government restricts overseas investments, rare metal speculation provides an alternative to real estate and stock market investments. If run well, these markets could bring light to backroom rare metal deals and confidence to businesses that they can secure long-term supply at a fair price. If run poorly, these markets could cause huge swings in prices that would undermine the market and leave companies short of material.

The Fanya Metal Exchange, based in Kunming, the capital of China's remote southwest Yunnan Province, opened in 2011, and is emblematic of the new exchanges. It has been wildly successful in attracting investors, despite resembling Arkansas in the United States, far from mining and farther from investors. In just three years since it started trading indium, it has amassed around four times the amount of indium the world consumes in one year. Fanya turned China, the world's largest supplier of indium, from a net indium exporter into an importer, as investors flocked to the market.[9]

This flurry of purchasing has old-time traders suspicious and has become a hot topic at international conferences for many reasons. First, price on the exchange had been nearly double what the market pays, raising fears that this is a get-rich-quick scheme with little basis on the real market. (For example, in May 2013, Fanya's indium price was over $900 per kilogram whereas the price physical traders used hovered around half that level, $540 per kilogram.[10]) Second, the market has functioned well in an environment of rising prices; many traders feel that if the prices fall, it could trigger a wave of selling that would affect the price of indium traded in the market.

Third, whereas most commodity exchanges are set up to manage the risk of trading between anonymous but creditworthy counterparties, Fanya's aims appear more nationalistic. "The goal of the exchange is to increase the value of China's minor metals industry chain and to enhance China's price negotiating power in the international market by using China's resource advantages and the new and advanced electronic trading tool," said Yang Guohong, Fanya's vice president, in Metal-Pages in 2013.[11] His comments echo goals by Jiang Yang, vice chairman of the China Securities Regulatory Commission, who commented four years earlier that future trading may help Chinese commodity importers "get fairer deals" as part of a "long-term goal of increasing our [Chinese] influence in terms of pricing."[12]

Such goals give rise to market fears that the government may be manipulating investments to achieve other domestic goals. For example, indium's increasing prices have helped to reduce the sales of illegal and legal metals abroad. "Smuggling has not been a problem since Fanya got going," an industry source told Metal-Pages two years after the exchange opened up. The smugglers, "simply sell everything on the exchange. Smuggled material doesn't turn up any more."[13]

Selling metals to investors who warehouse the material waiting for higher prices has the potential to starve manufacturers, who could make use of the material. But what might be far more worrying is that the exchange doesn't have the metal it claims to have, which means that people are holding a paper receipt for a supply that doesn't exist. This fraud is exactly what Chinese authorities uncovered at the Qingdao port in 2014. Some of the world's biggest banks were victims when Chinese trading companies took out loans using metals that did not exist for collateral. In 2014, the government estimated that companies had engaged in about $10 billion in fraudulent trade in just the first three quarters of the year, including the losses at Qingdao. Some banks fear that their losses might be only the beginning.[14]

Ming, a pseudonym I will use for a former executive who ran exchanges in China, has the confident air of a man who has seen all the machinations in China's commodity markets. He was part of a team that set up a trading exchange and was thus at the cutting edge of setting government policy that now encourages market-led, as opposed to state-directed, growth. He estimates that there has been an explosion of exchanges—between three hundred and one thousand different platforms for various products—over the past few years. He tells me that, despite this proliferation, the regulatory environment is weak and, "the real objective [of these exchanges] is to make money."[15] The exchanges are not geared to help provide accurate price information to the market—one of the main purposes of any market.

He attributes many of the challenges of the regulatory environment to the infighting and jockeying for power between government agencies regulating the exchanges. "The China Securities Regulatory Commission is so deeply involved in self-interested rule making, its rules make it impossible to set up an exchange for price discovery," Ming tells me. Part of his concern is that Chinese regulations haven't explicitly allowed for futures trading on most exchanges, despite the longtime roots of that practice in China. One of the main challenges in regulating Chinese exchanges is that they are governed more by policy than by strict laws and regulations. Unscrupulous traders responded by engaging in practices that push the bounds of those policies or they have disregarded them completely in hopes that policies would change to accommodate their practices. But even those who ran the exchanges pushed the regulatory limits.

To engage in futures trading, Ming tells me, he had to "tweak the trading model." He explains, "So if ten rules ban commodity futures contracts, you bend the rules so you only violate nine of them." Then you lobby the local authorities to prove you are not violating *all* the rules. In that way, you play off the existing power struggle between various agencies including the central government that want to cautiously expand trading with regulatory oversight and more ambitious local authorities.

Beijing has a right to be concerned with futures trading. Although futures contracts in China go back more than four thousand years to a time when people paid for rice delivery before planting season, just a few decades back the financial sector did not exist. The state controlled all banking activity and the trading of metals. The country's first large-scale experiments with commodity futures exchanges began in the late 1980s with a pilot program that quickly ran amok. Exchanges proliferated and soon there were fifteen formal commodity exchanges and another thirty less formal ones. Illegal trading overwhelmed authorities as the scale of trading grew and quickly expanded. It was, in essence, sanctioned gambling.[16]

Wang, again, a pseudonym of a trader from the time, tells me that all of his friends with whom he traded commodities in the 1990s at the Zhengzhou Commodity Exchange went to jail. He says that back then large institutions and the sons of high officials cornered certain markets through illicit and deceitful practices. Exchange officials would guide prices so that specific investors would profit. "This was completely the strong preying on the weak according to the laws of the jungle, a situation full of bandits and robbers." The practices became so egregious that the deputy president of the Supreme People's Court eventually referred to the Zhengzhou Commodity Exchange as a "big litigation family" because of the torrent of legal action engendered by unscrupulous trading.[17]

By the mid-1990s, authorities in China began shutting down nearly all futures trading along with many of the exchanges. It took a decade before the country began to gradually allow trading on government-backed exchanges. Wang blames the regulators for this situation, including the China Securities Regulatory Commission and the People's Bank of China, whom he feels were inept and complicit. "Oversight was very poor, and this helped to disguise collusion between those institutions and some market participants. At the time the amounts paid in bribes were large." The practice is not over, Wang adds, "[the amounts are] nothing compared to those paid today."[18]

While a broad spectrum of commodities trading is once again proliferating in China, regulators still have concerns. As Jian Yang of the China Securities Regulatory Commission stated in 2013, "laws and regulations are not properly implemented and violations are frequent."[19] Such comments are deeply disturbing when one considers the country's central role in the rare metals markets and its aspirations for more influence on metal pricing in the future. "Chinese people gamble," Ming tells me. "When you have a casino, people flock there." It's a tendency Ming took advantage of. "Our trading rules were designed like casinos."[20]

In 2012, Hong Kong Exchanges and Clearing Limited bought the London Metal Exchange, the world's leading metal exchange which provides a forum to buy and sell base metals as well as the minor metals cobalt and molybdenum. At a sale price of $2.2 billion, bankers at Morgan Stanley summed up market sentiment when they called the deal "extremely expensive." Hong Kong's bid was 180 times the net earnings of the exchange; by comparison, the previous most expensive deal for a comparable exchange was 66 times net income when the Chicago Mercantile Exchange bought the Chicago Board of Trade in 2007.[21]

The new buyers indicated that by moving the exchange to Hong Kong, because of its proximity to China, it would bring in new Chinese traders. More Chinese traders over the long term would make the exchange far more profitable, the argument goes. It will take time to hash out the new exchange's Asian business model; up to 2015, the results have been less than stellar. But profitability might not have been the owner's sole goal.[22]

When you run an exchange, you are privy to a lot of market information—the price of trades and volumes. A savvy data cruncher can learn much more: trading patterns, positions, and timing. Usually this information remains at the exchange, but Ming and some others are concerned that Beijing is using this information to the country's advantage.

"What happens in board meetings in Hong Kong this minute will be at the PBOC the next," Ming tells me, referring to the People's Bank of China. He fears that increasingly closer relations between Hong Kong and China—as exemplified by the Chinese decision in 2014 to pick the slate of candidates that will lead the island rather than letting Hong Kong residents pick their leaders—will wear down the wall between government and industry. And this in turn will turn corporate information into state intelligence. "There is absolutely no firewall," Ming tells me. And in a country where the line between the state and its largest companies is blurry, he is concerned that this information could give Chinese companies a leg up. "It [The purchase] was a geopolitical play," Ming explains.

Among my spam e-mails from Christianmingle.com and from financial firms touting reverse mortgages are e-mails from Chinese metal-processing companies. The sender—Della or Alice, usually a nominally Anglicized name, rarely with a last name—dismisses proper spacing between paragraphs, correct sentence structure, and capitalization rules. One e-mail sender even forgot her alias; the e-mail was from Alice but the note's first line began, "Hello I'm Candice."

The e-mails are similar—they boast timely, cheap deliveries of rare earth material or other rare metal products— employing the same sales tactics as those ads peddling erectile dysfunction pills. This is what separates rare metals producers from larger players like Glencore, the commodity trading giant. It doesn't spam potential customers.

In the 1980s and early 1990s, Chinese officials pushed to develop the country's resource base, from rare earths to indium, to meet growing domestic demand. But as soon as production capacity outpaced demand and flooded some global markets, the country introduced export controls such as quotas, taxes, and minimum prices. Disparities developed between lower domestically priced minerals and higher internationally priced ones. By creating a comfortable profit margin for illegal traders, this encouraged smuggling. Government crackdowns to reduce illegal rare earth trade have produced mixed results.[23]

For example, Chinese officials estimated that in 2011, traders shipped out 20 percent more rare earths illegally than legally. And these illicit traders continue to actively search for new clients. Few of the companies that spammed me seem to have export allowances under Chinese government quota systems. But as one woman I met at a trade show assured me in industry parlance, "we have quota."[24]

Smuggling is not just an anomaly of the rare earth sector; in fact, many rare metals have a history of evading China's production and export controls—from antimony, a critical element for fire-resistant products, to indium, discussed earlier and found in flat-screen televisions. These smuggled resources are sometimes so common that industry journals mention their prices.

Illegally mined rare metals in China are not easy to recognize. In warehouses in Vietnam, for example, they look much like their legally mined and processed brethren. In fact, illegally smuggled rare earths from China are quite legal in Vietnam, unlike, say, elephant tusks, which are illegal nearly everywhere. Inside the country they are discouraged, outside they are encouraged.

Material evades export controls in multiple ways. Under the protection of corrupt officials, illicit material leaves the country for neighboring territories like Hong Kong and Vietnam before being shipped onward. At other times, traders tell me, companies list rare earth powder as talc powder or other material on a ship's manifest but transforms into cerium powder on custom forms when the customer receives it. Smuggling is one of the main reasons why export and import figures of countries never seem to line up and which makes it challenging to understand the flow of materials.

"I know my customers buy stuff that has 'fallen off the truck' all the time through Vietnam or [has] been labeled something coming in and something else going out," says David O'Brock, the Estonian rare earth processor, referring to the way smugglers mislabel materials to evade controls. He says that the flow is stable. As he points to the white powder of rare earth oxides at his plant, he says, "This could be powdered milk."

China has reduced its export taxes on several metals, including indium, from 15 percent in 2009 to 2 percent in 2014, as part of a strategy to reduce the incentive for smugglers. As one trader commented, "This is good news as lower export duties will narrow the price gap between us and smugglers." But illegal trading is endemic in China.[25]

Despite Beijing's lip service to the vagaries of clandestine trading, many businesses and local governments benefit from it. Local officials who have been under pressure to sustain growth are loath to shut down profitable industries like mining and metal processing, especially because some media reports indicate that they may personally benefit from it.[26] But China is not alone in attempting to tackle illicit flows of rare metals.

"I'm an investor," Yudi, a trader of illegally mined minerals, tells me as we sip coffee at a café in Pangkal Pinang, a town on the Indonesian Island of Bangka, where a third of the world's tin is mined. Yudi, which is also a pseudonym, is in his late twenties, and sports a black-and-white plaid shirt. He has a long fingernail on his left pinky, a sign of wealth. His confidence belies his age.

"When I started three years ago, I was making 30 million [$3,000] a week." He shuffles his two BlackBerrys as he contemplates his words. "But more competition came and I'm making about that in a month now." And $3,000 a month is a huge bump up from his previous salary of $150 as a part-time travel agent. In a country where more than 100 million of the country's 250 million people live on less than $2 a day, he's doing quite well.[27]

His job is simple: he hires people to buy cassiterites, the main ore of tin, from illegal miners and sell them to local smelters. Cassiterites are the primary source of tin and secondary source of other metals. There are many other people like Yudi; some, mostly Chinese, also trade monazite. Unlike the tin ore, monazite can't be refined on the island, so it's helpful to have international connections. It's an open secret that some of Bangka's minerals containing rare earths—as well as materials from elsewhere in Southeast Asia—find their way to China, often through Singapore, skirting Indonesian laws that limit the export of unrefined minerals.[28]

Illegal trade in minerals is risky work. Yudi tells me that at the beginning of a workweek, an investor from Jakarta, Indonesia's capital, or China, will give him $100,000. (Yudi believes that many of these investors are electronic manufacturers.) He will then spend the cash to buy cassiterites from miners and traders, and within a week he will sell the minerals to a local refiner. Typically, he says, he is able to gain a 10 percent return, giving back $110,000 to his investors by week's end. "The investor does not care about how I get the minerals, I just need to deliver it in seven days," he tells me. There is no contract between Yudi and the investor; the deal is based on trust, a very limited commodity on Bangka. But Yudi takes precautions because he is on the hook to some powerful people.

Yudi has a few rules to reduce risk: never hold cash and make sure to have police support if something goes wrong. He researches the backgrounds of potential mineral subcontractors, who are responsible for buying the tin from individual miners, by asking neighbors and other business associates about them. "A little kid can give money to anyone; I need someone who knows tin," he tells me, illustrating the importance of having knowledgeable subcontractors. Unlike his investors, he requires subcontractors to sign contracts. He also pays off the police by giving cigarette money to the "little fishes," those newer to the force. For people higher up the chain of command, bigger sums are needed. He sends them payoffs via bank transfers or money delivered in person in manila envelopes. "The police work for me," he boasts. It is helpful to have such legal support and protection. One subcontractor, whom he refers to as a collector, absconded with his money; Yudi says he now owns the deed to that collector's house.

Yudi also pays off scientists at the smelters with $100 or so to ensure that they appropriately validate the grade of material he sells—a low grade means less money. About 70 percent of his revenue goes to his salary, 30 percent to payoffs or "fees," uncannily similar to corporate taxes in many parts of the world. The real risk to people like Yudi and the market is that if the authorities get serious about cracking down on his activities, metal prices will rise and Yudi will be out of a job. But that situation seemed highly unlikely during my 2013 visit to Bangka because the challenges involved in stopping illegal trading are enormous.

The money Yudi makes may be small by Western standards, but the money made in the entire illegal network is spread throughout the island—about 30–40 percent of the residents are connected to the mineral trade—and this includes some very important people, as he tells me, making a lot of money. Yudi says that even his friends are now making enough to invest in houses around Indonesia. With little sarcasm, he tells me that the market is corrupt. He is tired of dealing with most of the traders because they are "cheats." Yudi wants the market to change. "I want transparency."[29]

Despite the irony of wanting more rules in his illegal trading operation, he is in agreement with most other industry analysts who believe that the market would function more efficiently if the people working in it knew what was going on. The market for many minor metals becomes even murkier after Yudi gets the minerals to the refiner.

After tin and many other metals go through the smelting process, there is a fair amount of waste or "slag." Think of it as the seeds and rinds of an orange after you make juice. This waste has value, not in Bangka where facilities to refine it further are limited, but offshore. In 2010, Toyota Tsusho, the trading arm of the Toyota Group, announced that it would buy slag, with the intention of building a smelting facility on the island of Bangka to process rare earths. Now, the Indonesian government is pushing the local state-owned tin mine to produce rare earths.[30]

What's most discouraging from a resource standpoint is that many miners and refiners waste rare metals. Because not all minerals can be easily traded, either miners toss back minerals or processors throw away the sludge that comes from refining tin and other materials. As a result, precious resources are left behind, which necessitates additional mining operations elsewhere. Illegal and inefficient markets undermine the market by exploiting only the resources that can be easily traded and wasting the rest.

Unlike the proceeds from mining in Bangka that line the pockets of wealthy Indonesians and Chinese, in parts of East Africa and elsewhere, the money from trading minerals funds conflict. In the Congo, militias trade minerals containing gold, tungsten, tantalum, and tin.[31] In fact, about a quarter of all tantalum produced in 2011, came from Congo mines, but as Patrick Stratton, a tantalum expert at the consultancy Roskill points out, nobody knows exactly how much conflict tantalum is produced.[32]

The route these minerals generally travel from the mine is easy to track. It begins in hollowed-out areas of dense jungle where men on tiered hillsides dig up a thick, rich muck of dirt or outside rocky ferret holes where men with flashlights tied to their heads hammer away in the humid darkness. Traders bring sacks of minerals from these mines in Congo's western border areas on bikes or in rickety cars to main regional cities like Bunia. From there, traders bring the material by truck through Tanzania, Rwanda, or Uganda, where they sell it to another trader or ship it directly to one of approximately ten tantalum refineries, almost all in China. Once the material is there, smelters refine it. Then it is sold to another trader who will refine it further. Or it is sold to a component manufacturer who will add the material to specialty alloys that will eventually find their way into parts of flat screens or car airbag systems.[33]

In Colombia, FARC rebels, who have been fighting an insurgency against the government since 1987 and are labeled a terrorist organization by the European Union and the United States, trade tungsten mined from the depths of a national park in the Amazon jungle.[34] The traders haul bags of crushed rocks up the Inírida River through rapids so strong that they have to drag their long, narrow wooden boats along the shore. After a near weeklong journey through the jungle, the FARC traders transfer the crushed rocks to other traders who bring the bags by truck to Bogotá. Once in Bogotá, trading firms sell the rocks to international markets where traders and processors turn them into tungsten. The alloys or powders they produce have found their way into components used by BMW, Hewlett-Packard, and Samsung Electronics. After an article by *Bloomberg News* in 2013 about the Colombian trade, numerous traders stopped buying these materials. However, until the Colombian government intervened to halt this questionable practice, these materials continued to find their way to suppliers.[35]

Foreign governments, many nonprofit groups, and even companies themselves want to stem the use of these illicit materials. They see the proceeds from these minerals spurring human rights abuses by providing cash to armed groups that are in conflict. But the challenge is that modern supply chains are so long that companies don't know where their resources come from. For example, Philips Corporation has over ten thousand suppliers and is seven or more suppliers removed from the mines. Adding further complexity is the fact that most companies have hundreds if not thousands of components in their products. "A lot of companies have no idea whether or not they're using conflict minerals," notes Lisa Reisman, publisher of MetalMiner and a specialist on minerals from conflict areas. Indeed, even Apple notes that it does not have enough information to conclusively determine which country the minerals it uses come from.[36]

Tracing the route of commercial product components back to their source is far more challenging than most government officials in Washington or Brussels realize. "Policymakers have this bad misconception that a company has transparency, that they know what is in their products," says Randy Kirchain, a material research scientist at the Massachusetts Institute of Technology (MIT). "They don't know because of how much the manufacturing world has changed over the past several decades."[37] At one time a car manufacturer made most of its components, but today's companies outsource. As Kirchain tells me, Ford will specify the sound quality and size of a radio for their vehicles, but does not state what elements must be in the radio microprocessors that affect its function. Even Bas van Abel, the founder of Fairphone, a company specifically set up "to use only source materials extracted in acceptable humane and environmental conditions," doesn't know the origins of all his materials. And he may never know.[38]

In 2010, the U.S. government passed legislation forcing companies to state whether the materials they use in their products come from mines run by warlords in the Congo. The European Union (EU) followed suit with a watered-down proposal in 2014. The laws have little teeth because no one is banned for using materials from conflict areas and the EU's program is voluntary. But because large corporations do not want customers to think the products they manufacture help to fuel rape and murder, many multinationals are forcing traders to buy materials elsewhere.[39]

Consequently, the value of Congo-produced minerals is decreasing, which makes these cheaper materials available to companies that are less concerned about human rights. And that is a large number of companies, especially in Asia. For no-name brands serving a market with limited discretionary income, human rights may not rank high on their list of purchasing criteria. Companies serving those markets have greater latitude in sourcing the cheapest materials. And that material is easy to find. One trader told me that he had an offer to buy substantial volumes of cassiterites from Uganda—but Uganda has no significant mining operations, although it shares a porous border with Congo.[40]

The U.S. legislation also requires companies to report only newly mined sources of conflict metals. An increasing amount of scrap tantalum is actually primary material that is made to look like scrap to circumvent the legislation. Although there are no exact figures for the amount of scrap entering the market in this manner, MetalMiner's Reisman estimates that the percentage for tantalum is "perhaps even 'high' double digits."[41]

In other cases, especially in Asia, unscrupulous traders and processors mix conflict minerals with legally mined ones to ensure that they have enough metals to keep profit margins. "I would bank heavily . . . that Chinese firms are 'skirting' the law where possible," Reisman commented to me. Just as one ounce of sewage can transform a gallon of drinking water into sewage, grains of illegally mined minerals mixed with legal ones contaminate the entire batch in the eyes of Toshiba, Samsung, and other corporations concerned about their image.

The truth is that the illegal flow of materials continues to spread. Reports indicate that illegal mining is already producing more revenue for traders than does the illicit drug trade in South America. In late 2013, the Chilean newspaper *El Mercurio* reported that the business in Peru's Amazon region is already 15 percent more profitable than the country's lucrative drug trade. And as the prices of these minerals rise, conflict metals will only proliferate. But for many suppliers the risk of using conflict minerals is the least of their concerns.[42]

Yosi Sheffi, a supply chain expert at MIT, explains that for the supply chain of a company to remain resilient, it must rely on many suppliers; each of those suppliers in turn relies on a bevy of its own suppliers so that a shortfall from one supplier will be made up by another. If you draw this schematically with the end user at the top and with each supplier layer underneath, the chart resembles a pyramid with a wide base of component manufacturers and mining companies supporting one computer manufacturer like Toshiba.[43]

But the wide base of this pyramid and the stability of the whole chain may be a fiction. Take, for example, Toshiba's reliance on rare earth magnets for computer hard drives, speakers, and almost all small motors in their high-tech goods. Toshiba has many sources for the manufacture of its magnets, but those manufacturers ultimately rely on just a few provinces in China for the rare earth alloys to produce the magnet. From Toshiba's perspective its rare earth supply chain resembles a pyramid with the company on top and a network of suppliers below, but the structure may be closer to a diamond insofar as Toshiba's suppliers, and in fact all manufacturers globally, have historically relied on just one ultimate source—China. Sheffi tells me that companies "discover the diamond when there is a disaster." What he means is that if there is a supply shortage, despite Toshiba's numerous suppliers, the company as well as its competitors have nowhere to turn for components because their suppliers all rely on the same source.

To reduce the risk associated with limited suppliers and to ensure access to reliable supplies of rare metals, some companies are signing deals directly with mining companies. Boeing stepped up its alliance with the world's largest titanium manufacturer, VSMPO-AVISMA, in late 2010 to ensure its supply. Toshiba signed a deal with Kazakhstan's Kazatomprom in 2011 to establish a joint venture to produce rare metals including rare earths and rhenium. Likewise, Mitsubishi, Daido Steel, and the U.S.-based rare earth mining company Molycorp teamed together to produce rare earth magnets. It's an expensive proposition to tie up money in order to secure supply-line arrangements.[44]

At the same time, mining companies are moving further downstream. The Molycorp–Daido–Mitsubishi deal augments Molycorp's ongoing strategy to produce magnets in addition to mining rare earths. These strategies are fraught with risk as companies are getting away from their core competencies— farmers are skilled at growing tomatoes but that does not mean they have the skills to open a pizzeria. Many companies still would prefer to leave the work with the traders.

Shigeo Nakamura is blasé about the machinations of the market. He recalls previous price bubbles of rare earth elements

like the annual arrival of the typhoon season—destructive
for some, but inevitable. During his years trading rare metals
he has also seen indium prices and rhenium prices jump as
much as twentyfold.[45]

This volatility can be a headache and lead to financial
challenges for end users like General Electric that were caught
short of rhenium in 2006 when prices spiked tenfold.[46] To
Nakamura and his fellow traders, volatility is just a part of
work—a profitable part. "We hope for hiccups; we pray for
them," Danny Lehrman of Hudson Metals tells me back in
New York. While shortages strike fear into the hearts of companies, they offer opportunities for larger profit margins to
traders.

Nakamura now sees opportunity. "The materials revolution has begun," he declares. And those who know how to get
these rare metals to where they are needed are set to do quite
well, especially as the world demand for these metals for high-
tech needs is about to soar as we shall now see.

VI

Tech Needs
The Electronification of Everything

The ancient Babylonians enjoyed having a clean mouth. They dug plaque and food remnants from the crevices between their molars with the frayed edge of a twig. Their practice, though rudimentary, heralded the emergence of dentistry. In the 1500s, the Chinese improved on the Babylonian twig, producing toothbrushes from carved bone and bamboo fastened with bristles from the necks of pigs. Four centuries later, product designers abandoned the pig bristles and bone in favor of plastic and nylon. In 1960 the brush became electric, when Broxo, a company founded in 1956, sold a spinning brush and marketed it as advanced dentistry. The electric-toothbrush market has grown so large that one company, Royal Philips Sonicare, boasts that its premium product is in twenty-two million bathrooms. Today's offerings include battery-powered brushes and app-enabled cleaners that collect data about your hygiene.[1]

While the Chinese and the Babylonians needed only materials from the local forest or pig, manufacturers like Royal

Philips need resources from the plains of northern China to the hills of South America for the metals they need for the dozens of components that go into their electric toothbrushes. Usually when people think of the global economy, they think of soccer balls from Pakistan, or T-shirts from Indonesia, but electric toothbrushes involve exponentially more resources than these simple goods and, therefore, come from a far greater number of places than we have visited in previous chapters.[2]

A toothbrush needs circuit boards dotted with materials of tantalum in a capacitor that helps it to store energy; it requires a neodymium, dysprosium, boron, and iron magnet with material coming from southern China to provide the power to spin brushes in excess of 31,000 strokes per minute; and it needs batteries made from nickel and cadmium or lithium. Supplying the thirty-five metals it needs to make the electric toothbrush takes an extensive minor metal supply chain: miners like Brazil's Companhia Brasileira de Metalurgia e Mineração (CBMM) to supply the metal; Estonia's Silmet to process it; and the Lehrmans from New York to provide the alloys to component manufacturers, who sell their wares to the toothbrush manufacturer. It is a web that spans six continents. Just these components alone travel through seven countries—China, Congo, Chile, Russia, Korea, Indonesia, and Turkey. And the amount of critical material needed to make just a few small components adds up. Ryan Castilloux of Adamas Intelligence estimates that over 500 tons of rare earth material alone is needed in trace amounts for batteries.[3]

Whereas the origins of the electric toothbrush hark back to pig bristles and bone, its technological roots, and in fact for most products of the electronic age, come to us from break-throughs in places like the lab of Texas Instruments.

Jack Kilby joined Texas Instruments as an electrical engineer in 1958, only a few weeks before the company cleared out for summer vacation. Because Kilby, who served in the Office of Strategic Services during World War II, had just joined the firm, he couldn't take time off. So when his colleagues left him, as a new employee, he didn't know exactly what to do. Because the company made transistors, resistors, and capacitors, Kilby thought he could spend time trying to make them more efficient.[4]

At that time hundreds, if not thousands, of transistors, diodes, rectifiers, and capacitors—the components that modulated power in electronics—had to be soldered together by hand. This manual production was costly, raised the potential that just one bad connection would ruin the system, and held back the development of new electronics because the circuitry's complexity began impeding the flow of electricity. Kilby wanted to create one integrated circuit so companies wouldn't have to solder disparate parts together. With everyone on summer holiday, he had time to dabble.[5]

The success of Kilby's idea hinged on choosing the right material. He needed a metal that conducted electricity, albeit poorly, and that would allow the integrated circuit to regulate the electricity flow. Kilby made quick work, selecting the rare metal germanium. Just one month after his colleagues returned from vacation, he had his prototype. It looked like a stick of chewing gum stuck on a square of stained brown glass with wires protruding from it. But it was ingenious. His work won him the Nobel Prize in Physics in 2000, and the microchip became the backbone of microprocessors and our electronic lives. This was one of the critical seeds of today's Rare Metal Age.[6]

Less than a decade later, Kilby's integrated circuit had become a mainstay in the electronic world and was at the heart of the most advanced desktop wizardry of the day, the calculator. Starting in the late 1960s and early 1970s, Texas Instruments, along with other companies, began manufacturing calculators that did little more than add, subtract, multiply, and divide. By today's standards, they were merely functional office tools—that is, until Sir Clive Sinclair came out with the elegantly simple and aesthetically pleasing Sinclair Executive. In 1973, *Design Journal* heralded Sir Clive Sinclair's 2.5-ounce calculator as "a neat combination of up-market styling and advanced electronics trickery [that] manages to create a slot in the market all to itself. It is at once a conversation piece, a rich man's plaything and a functional business machine."[7]

At the heart of the calculator was the Texas Instruments semiconductor, a descendant of Kilby's work, but packaged in sleek black casing, the size of today's iPhone. It sported a thin black translucent screen that showed the calculation results in a red LED display, which itself relied on the rare metal gallium.

At £79, or about $190 at the time, Sinclair's calculator was relatively cheap and set the wheels in motion for executive playtoys that bridge the divide between work, style, and pleasure. But it had limitations. The battery had a life of only a few hours and it had an unfortunate tendency to explode when powered on. Shortly after the calculator came out, a model combusted in a Soviet diplomat's shirt pocket, reportedly causing the Soviets to investigate whether he was the target of a sinister plot. These initial shortcomings did not impede the calculator's growth because newer models were less volatile and featured greater functionality.[8] By 1986, they had become so widely used, as Kenneth Cukier and Viktor Mayer-Schönberger point out in their book *Big Data*, that about 40 percent of the world's general-purpose computing was packed into those handheld number crunchers—more power than all personal computers on the planet at the time.[9]

Sinclair's calculator helped to create "the first bottomless market for cheap electronic gadgetry since the transistor radio," according to an industry analyst at the time. It was the intellectual forerunner of simple, sleek hyperfunctional gadgets that extended over time to the Sony Walkman and the iPad. Sinclair realized that the future of high-tech gadgets was compact, powerful, and stylish. In essence, Clive Sinclair was Steve Jobs, before Steve Jobs. While genius lay in the product design, it was an entire new set of rare metals, starting with the ones Kilby used, that helped bring forth new electronic creations of old ideas.

"A lot of devices that were previously not electronic now have electronic devices in them," Walter Alcorn of the Consumer Electronics Association tells me. These devices seemingly bring the inanimate to life.[10] Take, for example, a stuffed bear named Teddy Ruxpin. When Teddy hit store shelves in 1985, he quickly became the best-selling toy of the year. Instead of a soft fluffy filling, Teddy had a motor, a speaker, and cassette deck all wrapped in a plastic shell. What Teddy lacked in cuddle factor, he made up for in personality. Pop a cassette tape into him and his mouth moved, his eyes blinked, and his arms flailed. Even better, Teddy talked. For many children, it was just another best-selling toy; for those in the toy industry, it was a defining moment. Teddy was not full of the number of rare metals that proliferate in today's electronic gadgets, but he was the first commercially available animated doll. And with ten million Teddys eventually sold, he was a huge step forward in the electronification of our toys.[11]

Looking back between 1980 and 1989, only two of the best-selling toys during the Christmas season needed batteries—Teddy Ruxpin and Lazer Tag, a game with electric toy guns. Previously the big hits were card games like Trivial Pursuit or dolls like Cabbage Patch Kids. But within ten years, Santa's gift bag became a whole lot heavier and far more rare metal intensive. By the first decade of the new millennium, nearly all best-selling Christmas presents contained electronics. They weren't just simple toys like Teddy; these were toy systems such as the Xbox and iPad. They were fast becoming the world's most technically advanced products. Toys once had form and shape to mimic trucks and play kitchens, but they are increasingly all about functionality and interactivity— and therefore reliant on rare metals such as the rare earth element phosphors in the screen for color. Because of such toys, Amber Pietrobono of Fisher-Price noted in 2013, kids ages three to ten are the largest growing segment of tablet users and the first generation of the Rare Metal Age.[12]

What might be most telling about our demands is that our resource needs are converging. A generation ago the calculator and the Walkman, the adult toys, were far different than the teddy bear and G.I. Joe action figures of my youth. Today, children's and adults' toys are virtually the same. We all hover over a screen; high-tech gadgetry is now embedded in the next generation. It is just one of the reasons why minor metal use is proliferating.

In 1983, a few years before Teddy Ruxpin, Motorola introduced the first mobile phone at a list price of $3,995. The nearly one-kilogram brick was more a businessman's showpiece than a serious communication tool; call quality was poor and coverage was weak. But as Motorola and its competitors developed more internal components, many from rare metals, reception and sound quality improved. Thirty years later the mobile phone is now a rich concentration of rare materials—60 percent of the phone is made of metals or ceramics.[13]

Each metal has a crucial part to play, like a player on a basketball team; without a particular element, you lose an important role. The antenna requires titanium and boron; the transmitter, titanium and barium; the condensers, tantalum and strontium; the speaker and microphone, samarium and cobalt; the connector, beryllium; and the power amplifier, gallium. Take out the rare earth element phosphors and the screen gets hazy. Take out tantalum and the phone gets bigger and will drop more calls. Take out gallium and the reception worsens.[14]

The reason the phones use so many metals is that the cost of each individual metal per product is so low, and their properties are so useful there is little cost consideration to switch them. And the low cost and high performance is only increasing as we develop newer technologies.

The evolution from cell phones to smartphones is dialing up our rare metal usage, especially considering that an estimated 1.5 billion of them will exist by 2017. Smartphones contain more metals, in greater amounts and often at higher grades than their predecessors. For example, 4g smartphones use six to ten times more gallium than a regular cell phone just several years before. The material use may be small per product, but it adds up in these small metal markets. One mobile phone uses on average just six grams of cobalt in its battery. The amount appears insignificant, but it equates to the consumption of about 7,500 tons of cobalt annually just for smartphones.[15]

To be sure, some new products use less rare metals than their previous iterations. For example, LED displays use far fewer rare earth elements per lamp than their fluorescent cousins. In laptops, new solid-state drives—which store data on flash memory chips—are replacing hard drives. Because hard drives use two rare earth magnets, one to help spin magnetic-coated metal platters and the other to encode data on them, the switch to flash drives appears to reduce the demand for the 10,000 tons of rare earth magnets we use annually to store pictures and files.[16]

However, while flash drives are faster and smaller, in 2014 they cost nearly eight times more than hard drives for the same amount of memory. Consequently, companies such as Acer are building computers like the Chromebook with lower amounts of flash memory. To offset this smaller memory, people are turning to cloud storage, where hard disk drives, with their rare earth magnets, are the backbone of remote storage. So whereas we are seeing a reduction in rare earths used in laptops, we are witnessing an explosion in rare earth magnets used in hard drives in cloud data storage centers.[17]

Rare metals help to make products smaller and lighter, as Clive Sinclair showed, but also to make them more powerful, thinner, brighter, stronger, and in many cases cheaper. And the better these consumer electronics become, the more we find them everywhere and begin to take their transformative properties for granted.

On a recent trip from Shanghai, my friend lamented the lack of a personal seatback screen on her flight. She squinted to watch a movie on the cabin's main TV screen a few rows ahead. Her experience without a seatback screen is increasingly rare on long-haul flights. Mary Kirby, editor of *Airline Passenger Experience Association* magazine, notes that it's a necessity for all long-haul airlines to order planes with individual seatback screens; without them, "it's almost like buying a car without a radio."[18]

We forget that screens were once just the domain of televisions. We have now become accustomed to screens or touch screens in subways, taxis, and even embedded in urinals. One of the main technical challenges to developing flat-screen technology was to find a material that would allow small light cells behind the screen to shine through but continue to conduct electricity, thus forming a complete circuit (and reducing the need for a bulky cathode ray tube).[19] The problem is that conductive metals are opaque, so putting a metal in front of the light cells blackens the screen.

Enter indium powder, which, when mixed with tin, becomes a unique transparent conductor that also sticks well to glass. Harnessing the material's conductive and transparent properties has made it essential to the development of flat-screen technology. But not much indium is needed for a seatback screen on an airplane—even a forty-two-inch flat-screen television contains just $3 worth of indium, just a sprinkling, a third less than the amount in a smartphone. (In fact, the indium-tin coating works so well as a clear conductor, a glaze of it now keeps airplane windows defogged and prevents frost buildup on supermarket freezer doors.) The metal's properties have been so unique that no commercial replacements have yet to unseat its dominant role in flat screens, despite shortcomings like its brittleness and inflexibility—properties that have contributed to many broken smartphone screens. And as the screens are becoming ubiquitous—they are expected to increase fourfold between 2013 and 2017—so will indium.[20]

Today's smartphones, tablets, and computers, which rely on flat screens, are now fundamentally different from the stand-alone products like the Sinclair Executive. The roughly 2.6 billion of them produced in 2015—a staggering number considering that the world population was 2.6 billion people just sixty-five years ago—function best when they are connected to each other. And it is this hidden network, the backbone of the digital age, that is also home to rare metals.[21]

The entire wireless network of phone base stations and relays could only have been built because of material science advances in the 1970s using metals such as indium, titanium, and tantalum. Critical materials like ceramics, made from barium and titanium, allowed base stations to relay telephone signals from your phone to the network.[22]

Now optical fibers, which are glass strings that transfer encoded light signals, are at the heart of the Internet as they move information around the globe literally at the speed of light. The key to successful fiber optics is to keep the light moving forward; this is a challenge because as light expands, it loses its intensity. Among the numerous rare metals in our fiber optic network is germanium. Manufacturers coat optical fibers with germanium tetrachloride to form a microns-thin seal around the core of the fiber. This rare metal functions as part grease and part insulator; it helps direct the light forward and prevents it from seeping out. Germanium makes up about 4 percent of the optical wire by weight, but this seemingly insignificant application has a big influence on a small minor metal market like germanium—about 40 tons of the 130 tons of germanium produced annually goes to keep the Internet humming. However, the market is set to explode; the EU estimates that by 2030 the fiber optic market will grow eightfold.[23]

In fact the demand for interconnectivity is so large that the International Energy Agency predicts that our information technology, communication, and consumer electronics will double by 2022 and increase threefold in fifteen years. And nowhere is this growth faster than in the Asia-Pacific where the rise of the middle class and its demand for these goods has grown about 20 percent in 2014. This seemingly unbridled growth may upend minor metal markets.[24]

Sunny is typical of the younger Chinese generation. When we meet in 2011, she extols the joys of her brand new iPhone4 as we talk over a hamburger at a hip new restaurant in Beijing's fashionable San Li Tun district where she lives. A fluent English speaker, she is a far cry from her parent's generation, children of the Cultural Revolution, who were taught to banish foreign concepts. Despite Beijing's slow Internet speed and the lack of Facebook or YouTube, she couldn't be more excited. "I waited all night for this phone," she tells me. It's hard to imagine this stylishly dressed woman braving an entire night on the sidewalk waiting to claim a phone.

Buying a new phone wasn't just a challenge in perseverance for Sunny, it was dangerous. At Apple's flagship store in Beijing, brawls broke out during the release of iPad 2 in 2011, and again with the iPhone 4s release in 2012.[25] But others endured far worse. In China's Anhui Province, Wang Shangkun, an eighteen-year-old with dreams of a better life, used the proceeds from hawking his kidney to buy an iPad and iPhone, but he ended up in renal failure.[26]

Whereas selling organs and even waiting outside all night for a store to open are extreme examples of what people will do to get the gadget they covet, they illustrate how the demand for high-tech goods knows few economic borders; this highlights a potential concern at the dawn of the Rare Metal Age: what if people in developing countries like Indonesia don't merely imitate the West's penchant for gadgets and connectivity—suppose they well exceed it?

Sesa-Opas, who prefers that I use his Twitter handle to maintain a thin veil of anonymity, tells me he won't tweet while riding his scooter around Jakarta. But the thirty-year-old does tweet behind the wheel, on the treadmill—wherever and whenever he has a free moment. Although he uses just a few apps—mostly Twitter, Facebook, and PATH, a popular message-sharing app in Indonesia—he estimates that he spends more than half his waking hours on social media.

Sesa-Opas's statistics prove his dedication. Over the past five years, he has sent over 142,000 tweets. That's more than 500 a week; about 74 tweets a day, every day for half a decade. "Twitter is like watching TV," he texts me. "There's news, drama and discussions everywhere. If you follow the right people, it's non-stop entertainment on your mobile phone." Every morning after turning off his smartphone alarm, he checks his Twitter mentions. "I'm addicted," he tells me. He has not been away from his "TV" for more than 36 hours in five years. When a work trip brought him to a remote part of the country, he was without Twitter. "I was VERY anxious for the first 3 to 4 hours," he says. "When I logged in later, it seems I missed all the fun." It took him hours just to get caught up.

He's not alone. Throughout cafés in Jakarta, young Indonesians sit three, four, and six at a table, with iPhones or Samsungs in one hand and tepid sweet tea in the other. It's common to see entire families out together, each person staring at or swiping on a handheld screen. Indonesia has the world's second largest Facebook community and it is home to the third most active users of Twitter. There are more active tweeters in Jakarta than in New York or Tokyo according to a Semiocast study during a random month in 2012. Even Bandung, an Indonesian city of 2.5 million, had more tweets than Paris, Los Angeles, or Chicago. This usage influences the global conversation. When the Indonesian personality Eyang Subur divorced four of his eight wives, the news was trending globally on Twitter, prompting a confused Emma Watson, an English actress, to tweet to her six million followers, "Seriously, I mean who is that guy?"[27]

Indonesians' strong Internet presence is all the more astounding because only a quarter of the country's nearly 250 million people have access to the Web. Many cannot afford gadgets to get online, and for some, the infrastructure such as fiber optics and phone towers has not yet reached them. That is changing. In 2013, the number of mobile data subscribers nearly doubled over the course of the year to 58.7 million. All told, there are more mobile phones than people in the archipelago, with about 1.2 phone subscriptions per every person. Ten years ago, while traveling in the cities, I was hard-pressed to find reliable Internet connections. Now nearly every café seems to offer a connection, with large numbers of people like Sesa nursing a coffee and hooked to their screens. And guessing by his past behavior, Sesa will be one of the first to buy wearable electronics; it is a sector that didn't exist a few years ago but is expected to lead to sales of more than 110 million products by 2018.[28]

All this usage poses a resource conundrum. Thomas Graedel at the Yale School of Forestry and Environmental Studies suggests that if resource demand grows in line with economic projections, the total metal flow needed to meet demand will be five to ten times greater than today's level.[29] That means we will need five to ten times the amount of raw material to meet the appetite of a growing and increasingly wealthy Indonesian and global population. The demand for resources, especially minor metal resources, may outpace even Graedel's amazing expectations. But it's not only personal products that rely on rare metals; the world's largest products do too.

Boeing's manufacturing facility outside Seattle is not just big— at 472,000,000 cubic feet, it is enormous, and dwarfs the Pentagon. The facility is so large that employees take shuttles from the parking lot to their offices. The building seems oversized until you see the assembly lines: rows of planes, the biggest high-tech products on earth, nose to tail. Perhaps no single product uses more minor metals than an airplane. If you don't think planes are high tech, consider that the wires connecting the electronic wizardry in just one Boeing 747 extend over 130 miles. But the electronics onboard are just the beginning. Planes over the past fifty years have shed steel, replacing it with composite materials and lighter metals like titanium. For example, the new Airbus A350 frame is 14 percent titanium compared to 6 percent in the older Airbus A320. The increasing use of titanium is expected to lead to a doubling of titanium's use in just five years to 41,200 tons by 2016.[30]

The most complex materials are increasingly found in engines. The National Academy of Sciences notes in its report, *Minerals, Critical Minerals, and the U.S. Economy,* that the recent advancements in propulsion systems are only possible because of improvements in using the high-temperature properties of minor metals like cobalt, rhenium, and yttrium. For example, rhenium, because of its strength and higher melting point, allows jet engines to run hotter and consume less fuel.[31]

But as a by-product metal of copper production, rhenium is more scarce than gold. Processors can only squeeze one ounce of rhenium from more than 120 tons of copper ore—that is like extracting one ounce from the weight of forty-five SUVs.

Between 2006 and 2008, the price of rhenium jumped eleven-fold, to more than $11,000 a kilogram as the demand increased far faster than processors could increase the supply. However, it is easier for engine designers to redesign the entire engine to run at lower temperatures than it is to find another metal to replace rhenium.[32]

General Electric's (GE's) aviation business was short of rhenium in 2006. Because of its troubles in buying the metal, the company instituted a rhenium reduction program, which included harvesting the metal from old engines, capturing metal grindings lost in manufacturing, and trying to use less material in its designs. Each turbine blade needs just a dusting of rhenium in its metal alloy, about half an ounce (or 14 grams). While GE's strategy has reduced the risk of a rhenium shortage, such efforts won't work for all metals.[33]

"Each individual element has its own strategy," cautions Steve Duclos, chief scientist at General Electric Global Research Center. There is no one-size-fits-all approach. He adds that it's not so difficult to save 30 percent in the use of a particular element because using less material optimizes manufacturing processes. But companies often run low on options after becoming more efficient.[34]

GE has increased the security of its rhenium supply with its recovery and recycling program, but only to a point. The company still has to buy some rhenium on the market. And because not every company using rhenium is as forward-thinking as GE, shortages can nevertheless force prices up or cause manufacturing delays. What may be more worrisome for GE is that the company's efficiencies paradoxically have made it more vulnerable to rhenium shortages in the long run. Because it now uses the metal so sparingly, it will become an even greater challenge to use less rhenium in the future.

Duclos notes that changing manufacturing processes can reduce demand, but replacing a rare metal with another material would be a nice quick fix. However, that is often daunting. "You can't take one out and put one in and say, 'here you go.' It's not as simple as people think," says Karl Gschneidner, the material scientist known as Mr. Rare Earth.[35] Searching for alternatives to replace rare metals has a substantial cost. "It takes away resources from developing new technologies," Duclos says. The process is also time consuming. It can take as much time—or even more time—to find suitable replacement materials as it did to find the original material, which can be a decade or two.[36]

The rhenium price spike also reveals something deeper: a resource trade-off at the heart of the Rare Metal Age. A substantial reason for the run-up in rhenium prices was related to the increase in oil prices just after 2001. Since fuel accounts for around a third of airline costs, airline executives wanted more fuel-efficient planes. And those plane engines needed rhenium; therefore, demand for it increased. In essence airlines were exchanging dependence on fuel for dependence on rhenium.

Boeing and GE must continuously understand the evolving rare metal trade because of their vast demand. Boeing's 747 needs six million components from over thirty countries. What's more, rhenium is just one metal of the seventy or so elements on the periodic table that General Electric and Boeing together use. Boeing likes to run a lean supply line; only five months elapse between the time that a component enters the manufacturing supply line until it leaves as part of a plane.[37] While not building buffer supply stocks means that the company has more cash to spend on more productive investments, running such a lean supply chain puts the company at risk in ensuring long-term supplies of rare metals. Boeing, General Electric, and other companies with long supply chains have beefed up rare metal risk management committees—some had met as often as once a week immediately after the rare earth crisis in 2010.[38]

Such rare metal risk committees are a relatively new phenomenon insofar as manufacturers have historically paid little attention to those supplies because they constituted a very low ratio of a product's final cost. Randy Kirchain, a principal research scientist at Massachusetts Institute of Technology (MIT), tells me, "Most companies have not been focusing on these resources; they were focusing on high valued resources—if you are an auto company, that would be steel."[39]

But CEOs have gradually become concerned, according to a study by the consultancy PwC entitled, "Minerals and Metals Scarcity in Manufacturing: It Is the Ticking Time Bomb." More than two-thirds of manufacturers fear resource scarcity will soon affect them. As Bill McClean, president of IC Insights, Inc., a consulting firm for the integrated circuit industry, notes, "A lack of a 10-cent capacitor can shut down the whole line for a $300 product." It is just such a tantalum capacitor that purportedly delayed the launch of PlayStation 2 in 2000; another tantalum shortage seven years later at Boeing allegedly contributed to delayed production of its 787s.[40]

Companies are often more beholden to a particular material than to a product. For example, a lithium-ion battery company cannot switch to making nickel metal hydride batteries because the production processes are so different. Merely switching suppliers causes headaches, as John Smith, a metal trader at 5N Plus, tells me, because they must test the quality of material over a long period each time. Even slight variations can throw off a company's manufacturing process.[41]

Despite these tremendous risks, many companies still don't know all that is in their products because of the complexity of sourcing. For example, MIT's Randy Kirchain explains that Ford Motor Company outsources specifications for its radios and then selects the one that performs best.[42] The company spends little time worrying about the rare metals in those radios. That mindset is slowly changing. "Companies are waking up . . . as their tech portfolio becomes more diverse, they are tapping into a much broader set of elements," he says. "People can absolutely get caught in not having enough material."

"Companies get dumb quick," Kevin Moore, a General Motors veteran of more than thirty years, tells me.[43] Once a crisis passes, they move on and revert to old ways, which means spending less on low-probability, high-risk events, like rhenium shortages. But that's exactly when companies should pay attention to supply security. The problem is that when the price of these materials falls, so does executive interest. The challenge as GE's Steve Duclos noted at a symposium at the Materials Research Society conference in November 2012, is that "Crises are more costly, more frequent and more distracting."

The crux of the problem is that consumers are demanding new high-tech products and technologies from companies like Apple and GE far more quickly than mines can be developed to increase supplies. Basic economics teaches us that the higher price of a resource leads to more production, but this does not always work for rare metal markets. For many rare metals, supply does not move in response to demand, but price does because companies cannot quickly produce more. Mining companies cannot set up a mine to produce dysprosium in response to a growing need next year.

This volatility makes diligent monitoring of minor metal supply lines crucial. As more products follow the electronification path of the toothbrush and more people use electric toothbrushes, the potential for rare metal shortages increases. Companies can only do so much to insulate themselves. They are often at the whim of market forces. Duclos notes, "We can't solve all the problems on the periodic table." Ensuring the supply of rare metals for every use is just not possible. And although the high-tech demand for rare metals is daunting, it is dwarfed by demand for green technology, which is set to soar.[44]

VII
Environmental Needs
Rare Metals Are Green

There is little to see along the 115-mile (185-kilometer) road from Sillamäe to Estonia's capital, Tallinn, early on a January morning. At 8:00 a.m., the sun hasn't yet risen, and, even if it had, Estonia's flat expanse of white snow changes little along the ride. The homogeny of the natural scenery conceals a resource buried beneath the frozen ground. Trapped within the soft-brown sedimentary rock is oil shale, which the Estonians have been mining for nearly a century. A cheap but dirty fuel, oil shale is still the source of 70 percent of the country's energy. Aboveground, the brisk, frigid morning air turns warm plumes of smoke, coming from an oil shale energy plant in the distance, into faint billowing clouds. Beyond these clouds, I soon see a promise of a cleaner future—several wind turbines, their blades turning in unison in the Arctic breeze.[1]

These new wind turbines provide about 10 percent of the country's total generation capacity. Ambitious plans being discussed in Tallinn promise an increase from some 300 megawatts (MW) of wind power to 1,800 MW in the near future. Like the power plants running on oil shale, these factories are making use of local resources: wind and neodymium, the metal produced at Estonia's rare earth element processing plant in Sillamäe. Although the basic energy source, wind, is free, we need costly metals to harness it.[2]

Green applications are far more than just wind turbines and solar panels; they are energy-efficient cars, lights, and even elevators. And just about all these technologies, from ocean tide turbines to battery packs, require rare metals in their infrastructure. But green technologies are more than the products. Some rare metals themselves should also be considered green because many of them, like niobium, drastically reduce the amount of other metals that are used, meaning a smaller overall carbon dioxide (CO_2) footprint. And as abhorrent as this may sound to some environmentalists, green goals require increased mining and more processing of rare metals. Mining is not antithetical to a green economy; it's a necessity.

And studies show we are going to need more of them—a lot more of them—to curb global warming. According to the UN Intergovernmental Panel on Climate Change, renewables must supply about 50 percent of the world's energy by 2050, thereby reducing the importance of fossil fuels in our energy mix. The study concludes that the world must nearly eliminate fossil fuel use by 2100. The road to change is not simply about switching to new technologies. It's about ensuring rare metal resources. In a joint study in 2011, the Materials Research Society and the American Physical Society issued a warning that should be of grave concern. It states, "A shortage of these 'energy critical elements' could significantly inhibit the adoption of otherwise game-changing energy technologies."[3] This means that we could be condemned to a fossil fuel world, if we cannot bolster the rare metal supply lines we need to support our green technologies.[4]

The International Energy Agency (IEA) predicts that to keep global warming to an increase of no more than 2 degrees Celsius, in twenty years renewable energy must generate half the electricity on the planet. To meet this goal, the IEA assumes in one of its likely scenarios that combined solar and wind power must produce over 6,000 terawatt (TWh) hours. That's an enormous increase compared to the combined 750 TWh that the world produced in 2013. What's more, car manufacturers must increase production of electric vehicles by 80 percent annually from eighty thousand produced to seven million so that at least twenty million will be on the road by 2020.[5]

If we could get the world to buy enough wind turbines and electric vehicles to blunt the effect of global warming, it would require a more than 700 percent increase in neodymium production and a 2,600 percent increase in dysprosium over the next twenty-five years, according to some research at the Massachusetts Institute of Technology (MIT). The challenge is that historically the production of these two rare earth elements has only increased at a rate of 6 percent per annum, not by the 8 percent or 14 percent needed over the next twenty-five years. Elisa Alonso, a former MIT researcher, who now advises agencies on natural resource security, says the real problem is that despite vast fluctuations in the production of particular minor metals during any given year, few metals have sustained such growth rates over a decade. Calculations by the U.S. Department of Energy and the European Union, like those by Alonso and her colleagues at MIT, show us short of the materials needed to produce the environmental change we require.[6]

The blades on 3 MW wind turbines spinning above the frozen highway to Tallinn rely on rare earth permanent magnets, part of a direct drive system, to help convert the spinning blades into electricity for thousands of Estonian homes. These magnets are an integral part of this relatively new breed of turbines, which were introduced commercially in 1992 and have become more widespread in the past decade. In China, they already make up roughly 25 percent of the market, a figure that has been increasing. In older technology systems, gearboxes in essence fulfill the role of helping to convert the speeding blades into electricity, but due to the tremendous mechanical stress on the gears, many lasted far fewer than the twenty years once predicted. One of the reasons for the increase in permanent magnet systems is the lower-than-anticipated maintenance costs because these systems use fewer parts.[7]

Less maintenance makes these newer magnets more attractive, particularly to offshore wind turbines that are difficult to service and must withstand harsher elements because of their location. The wind turbine manufacturer Alstom reports that a direct drive system that uses permanent magnets cuts maintenance costs that can be as high as 40 percent of the total cost for wind turbines.[8]

Unlike the stereo speakers or computer hard drives discussed in the previous chapter, which use rare earth magnets by the gram or half ounce, these hundred-meter-tall towers use rare earth magnets in amounts thousands of times more. One turbine alone can use anywhere from 250 to 600 kilograms of rare earth magnets per megawatt of capacity. So while the wind turbines on the ground in Estonia use close to a ton of rare earth magnets, the increasingly larger turbines, up to 10 MW contemplated for offshore, could use three times the amount of magnets. That means wind turbines are using around 2 tons of rare earth magnets each, which translates to usage of up to 160 kilograms of dysprosium, one of the most difficult rare earth elements to source.[9]

Dysprosium supplies present a challenge. A future hybrid or electric car will need less than a 100 grams and a future wind turbine about 30 kilograms. But with deployment of electric and hybrid vehicles in the millions and wind turbines in the thousands, a large increase in demand can be expected. According to the U.S. Department of Energy, we could need 8,000 tons annually, about seven times more than was produced in 2010.[10] These analyses did not even consider the exponential growth of dysprosium needed in high-tech goods.

Fear of resource shortages is giving companies pause as they select technology for their turbines. According to a study by PwC, a consultancy, almost 90 percent of business leaders in the renewable energy sector believe that mineral and metal scarcity will impact their businesses, and about 80 percent expect scarcity to become a larger concern in the future.[11]

Gareth Hatch, cofounder of Technology Metals Research, notes that several months before the rare earth crisis in 2010, magnets in some of the larger wind turbines cost about $80,000 per turbine. Within a year, as rare earth prices skyrocketed, the cost jumped to well over $500,000. Companies were scurrying to find rare earth supplies. But eighteen months later, prices returned to lower levels. Hatch says, "Companies still have a hangover from the price [spike], they are not eager to rely on these metals."[12]

"Companies want to reduce the risk in their supply chain," says Eize de Vries, a noted wind expert. At the height of the rare earth price spike in 2011 he commented that it seemed just a matter of time before the industry switched to turbines that relied on permanent magnets. But he also said that the rise in rare earth element prices led to a major increase in generator manufacturing costs, which pushed companies to rethink their drive systems. For example, Siemens is phasing out dysprosium in its turbine. And companies like General Electric (GE) and Vestas are developing their older, rare metal-free gearbox technologies. This is striking, especially for GE. In 2009, after acquiring ScanWind, a Norwegian firm with permanent magnet, direct drive technology, Victor Abate, GE's vice president of renewables, commented, "This acquisition will give GE the ability to provide a direct drive, offshore wind turbine offering as an option to our customers . . . we look forward to further developing their proven technology."[13]

The rare earth price and supply concerns of 2011 pushed the industry to explore further hybrid drive trains, which use roughly a third of the amount of rare earths as direct drive turbines do, and to consider other alternatives.[14] But green tech companies aren't just changing their turbine plans; some are trying to avoid certain rare metals altogether for fear that China or another country may make trade challenging.

Michael Silver, CEO of American Elements, believes that too many countries and companies are also shying away from using a particular minor metal because of geopolitics. They fear a repeat of the rare earth price shock in 2010, specifically that geopolitics will reduce market supply or drastically increase the price of a material.[15]

What concerns Silver is that companies are spending vast sums of money to switch away from reliable metals to untested alternatives that appear to face less geopolitical risk. Silver calls this decision to use a less-than-best option because of fears over geopolitics, "innovation distortion." And these geopolitical, not geological or manufacturing, considerations are leading companies to develop second-best technologies. Switching material to shore up your supply chain may make sense in some cases, but it provides the space for companies that don't share your supply chain concerns to develop better products because they continue to use the known rare metals. In addition to the market risks, substitution in and of itself is an area of increasing research.

Bill McCallum, a senior materials scientist, is on the front lines of research at his lab at Ames Laboratory in Iowa. He leads U.S. government efforts to reduce reliance on neodymium-dysprosium magnets. His goal is for U.S. companies to rely on homegrown resources for their technologies. With supplies of dysprosium limited, McCallum hopes to discover a way to replace it with cerium, a soft, silvery metal with the atomic number 58. Ductile and malleable, cerium is also the most abundant of rare earth metals. He wants to fundamentally reshape the auto industry and is looking for a second-best solution for one of the most vexing problems facing the United States—ensuring the development of green technologies.

Although cerium is cheaper and far more abundant than neodymium and dysprosium combined, it results in a weaker magnetic force and erodes quickly at high temperatures. Because McCallum uses a substitute material, his expectations are low, "we are trying to come up with a magnet that is not a very good rare earth magnet," he tells me.

So even if he does achieve a breakthrough, his magnet will never be as strong as its predecessor and manufacturers will have to redesign whatever system contains the magnet so as to compensate for the lower strength. McCallum, who receives support from the U.S. government as well as General Motors, hopes his magnet will end up not only in wind turbines but also in hybrid and electric vehicles because 90 percent of the market relies on permanent magnet motors. But he tells me he needs a lot of help. Every member of his team faces obstacles and needs at least "one miracle" to find a working substitute.[16]

If successful, his magnets would end up in products like cars. In terms of rare metal resource use, it's best not to think of a car as one product but as a house of appliances—stereos, air-bag systems, engines—all working together. A standard vehicle has more than forty magnets and twenty sensors that use rare earth elements weighing close to half a kilogram. New features, such as cameras that replace mirrors, only increase their sophistication and minor metal demand.[17]

What's most concerning to car companies now is that each hybrid vehicle uses up to 1.25 kilograms of rare earth magnetic material and each electric vehicle about three times more. Rare earth magnets are in electric cars for the same reason they are in wind turbines: they are far more efficient and powerful than the induction motor, in this case. "For the same amount of power output, compared to an induction motor, the permanent magnet motor is always going to be smaller, lighter and more compact," says John Miller, a former power electronics researcher at the Department of Energy's Oak Ridge National Laboratory. This poses a dilemma for car companies.[18]

"It's hard to move away [from permanent magnets]," says David Reeck, manager of electrification strategy at General Motors (GM) China. As there is only so much room under the hood and so much weight an electric or hybrid motor can push, minimizing mass and size are therefore critical. This means that most car companies have focused heavily on permanent magnet motors.[19]

Kevin Moore, who has served on GM's procurement team for thirty years, tells me that before 2010, GM, as well as many other car companies, had little idea of the precariousness of their rare earth metal supply line or how much each car needed. GM now knows. And that has changed the company's thinking.

Moore was part of a GM delegation sent to China to assess supply risk. It was a quick trip. "If you can't assure me the material, then I am not going to make the car," he recounts a lead engineer at GM having told him. No one could provide this assurance, and the delegation came away looking for alternatives. "If these rare earths were readily available, there would not even be a discussion as to what [motor] to use," Ed Becker tells me in early 2014. Becker spent an equally long time at GM and was responsible for reducing rare earths in magnets. He continues, "Other than supply, I can't think of another negative for these [permanent magnet] motors."[20]

Car companies need to secure supplies for the entire production cycle of their vehicles, which is nearly a decade—roughly three years from design to production and then a subsequent seven-year production cycle. It is difficult to make even small modifications, much less change an entire motor. As Pete Savagian, general director of Electrification Systems and Electric Drive Engineering at GM, notes in an interview in 2013, "We can't simply change overnight."[21]

Another concern for those like Savagian who are responsible for material supply is that green vehicles use the same minor metals as are used in other green and high-tech applications. Therefore, GM must assess the future demand and criticality of rare earths in other sectors such as wind power. If he finds that wind turbines manufacturers have no alternative to rare earth magnets and the demand is set to grow, in a time of low supply, GM will find itself in a costly bidding war. Referring to the rare earth market, Savagian says, "There is an underlying instability. . . . We must consider the costs and the volatility of the costs." Moore tells me that the company was so concerned about supply, that his colleagues even discussed buying a rare earth mine. Since then, GM's hesitancy about the future of rare earths has led them to review options for developing the less powerful induction motor.[22]

GM is not alone says Ed Becker: "Every motor company I know has a backup for a permanent magnet motor." Developing a motor using an older induction motor design that relies on copper or aluminum electromagnets instead of rare earth ones is a hedging strategy. Only a thin line separates hedging from innovation distortion. GM's decision about which motor to make is not made lightly; a wrong choice could put them out of business.[23]

In the late 1890s, a similar battle loomed between horseless carriages and propulsion systems. Steam, gas, and electric engines each had their strengths, but, at the time, gasoline-powered vehicles made up less than a quarter of the market. Ransom E. Olds, an engineer who eventually amassed thirty-four patents, had his roots in steam energy and steam-powered cars, which were less expensive at the time. Despite the cost difference, Olds believed that the future of car development was in the power and efficiency of a gas engine. Because of his foresight, we know the car named for him, the Oldsmobile, and not cars named for his competitors, Stanley or Pope.[24]

Some might argue that GM is now at a similar decision point. The hybrid and electric car market is expected to soar.

In ten years, sales of hybrid and electric vehicles will nearly quintuple to $334 billion from $69 billion in 2013. Choosing the right motor is critical not just for GM. It's critical for the U.S. economy, as Washington has shown by its unwillingness to let one of its biggest companies go under during the financial crisis.[25]

Despite the need for secure supplies of rare metals, manufacturers, like investors, prefer to put as little money as possible into ensuring supplies. The market has punished companies that took unnecessary precautions. In the late 1990s, the price of palladium, one of the platinum group metals used in Ford's catalytic converters, spiked roughly threefold when Russia, the world's largest supplier of palladium, held up shipments. Like all car companies, Ford used platinum group metals in its converters. The company originally relied on platinum, but switched to its cheaper sister element, palladium. These metals filter the exhaust and reduce car emissions by chemically converting three pollutants—unburned hydrocarbons, carbon monoxide, and nitrogen oxides—into less harmful exhaust.[26]

However, platinum group metals are some of the most expensive elements in the world, often more expensive than gold—so expensive that they are sold by the ounce. Although cars use only 3.5 grams each (about 11 percent of an ounce), it adds up when companies use 9 million ounces a year, according to an analyst at Ford Motor Company.[27]

Since the introduction of catalytic converters into cars in the 1970s, car companies have been trying to replace platinum group metals with other minor metals such as cobalt or iridium, but to no avail. Ford even reached out to China to develop catalytic material from rare earths in an attempt to trade the company's reliance on one group of rare metals for another. But the replacements never worked.[28]

By 2000, Ford's purchasing department was nervous about constantly rising palladium prices and thought it would be a good idea to secure vast amounts of palladium. The department's purchase seemed prudent as prices continued to rise to $1,082 an ounce in early 2001. But the price quickly fell to around $300 an ounce over the year. The U.S. company took a $1 billion hit, an estimated loss of $800 an ounce and an incalculable loss to its reputation. "These surprises over time begin to affect the credibility of the company," said Rod Lache, an analyst at Deutsche Banc Alex.Brown. Ford's loss led to a lawsuit. Such a loss also encourages companies not to stockpile rare metals and overly hedge their positions. It may have even encouraged them to avoid rare metals when possible.[29]

Economists will tell you that for buyers to freely accept hybrids or electric vehicles (EV), the performance of the vehicles must match that of the internal combustion engine, about $250/kWh, a measure of cost performance per unit of energy, at today's gas prices. Most EVs were at $485/kWh in 2013, dropping by nearly half in less than a decade. Tesla's Model S gained 8.4 percent of the luxury market in the first half of 2013, outperforming the established Mercedes S class and BMWs, and capturing the Motor Trend Car of the Year award, but at more than $72,000, it's out of the reach of most car buyers.[30]

Every few years over the past decade Jon Sonneborn and his business partner have bought the same car. Business has been good, so it is usually been a Mercedes or a Lexus, but last year it was a Tesla. Sonneborn tells me it is comfortable to drive, American-made, and goes from zero to 60 miles an hour in 5.5 seconds, although he assures me that he doesn't know this from personal experience. Plus, the car has no tailpipe and no direct emissions. "It feels good to be driving a green car," he says.[31]

His Tesla is already running at $250/kWh of storage capacity. Of course, the secret is rare metals. Although Tesla's induction motor does not rely on rare earth permanent magnets, many of the other car components rely on rare earth permanent magnets along with other rare metals such as titanium, to protect the large car-length lithium-ion battery, which alone costs around $30,000, one reason the vehicle costs so much.[32]

The consultancy McKinsey notes that with gasoline prices at about $3.50 a gallon, car companies that use batteries at prices below about $250 per kWh could produce electric vehicles competitively. But what if the next generation of Tesla becomes cheaper to manufacture? Tesla may transform not only the car industry but also the green energy market and the demand for rare metals. The company is now building a huge battery factory, a $5 billion behemoth, producing as much lithium-ion battery power as all other battery factories in the world combined. Tesla's goal is to drive down the cost of its battery by at least 30 percent and put the car within reach of those who admire the model, but cannot afford it. Reducing the cost of the battery and improving its efficiency is crucial. According to the Department of Energy, an electric vehicle that is cost-competitive with a gasoline-powered car would need a battery with double the energy storage of lithium batteries at 30 percent of the cost.[33]

Reducing the cost is essential to the electric car market, according to Cyrus Wadia, who advises the president on green energy and materials at the White House Office of Science and

Technology. High costs remain an obstacle to the widespread use of battery energy storage. But reducing costs to spur demand also stresses the rare material market insofar as there are few material substitutes for batteries.[34]

Electric car batteries for cars smaller than Tesla weigh more than 200 kilograms (about 440 pounds), cost between $8,000 and $18,000, and are often made from lithium.[35] Alex Teran, a research scientist at Leyden Energy, says that the reason the battery industry relies heavily on lithium is that no other element is lighter or more conductive than lithium. "It's proven its superiority. There are no surprises left; you can look at the Periodic Table and see what we can come up with."[36]

If Tesla meets its goal of reducing battery cost, it will transform not only the car market but also the lithium market and other rare metal markets. Some are skeptical that a low-cost, high-quality supply of the metals will be there. As Wadia noted in 2010, "Scale-up will require a long lead time, involve heavy capital investment in mining, and may require the extraction and processing of lower quality resources, which could drive extraction costs higher."[37]

Others like Stuart Burns at MetalMiner, a metal market intelligence firm, estimate that the sheer volume of the new battery facility will suck the cobalt market dry unless cobalt production quickly expands. Some, including Technology Metal Research's Gareth Hatch, fear it will do the same to the market for graphite, which Tesla will also need in large quantities. Gareth notes that the Tesla plant would need 102,900 tons of graphite, 125 percent of the total current global supply. Assuming Tesla can meet its goal, the new batteries will spur new demand in the automobile sector as well as in the green energy sector. And that's not the only battery in the market. Start-stop batteries, about the size of a shoebox, are coming into gas-powered cars because they save fuel by allowing the engine to turn off at stoplights and in traffic.[38]

Batteries are critical technologies for the deployment of solar and wind turbines. They are so critical in fact, according to Lyndon Rive, the CEO of Solarcity, that by 2020, his company might not be able to sell solar panel systems unless it packages them with batteries. The reason is simple—without sun, of course, the solar panel cannot produce energy. Batteries, which the solar panel charges when the sun is shining, can discharge their power when it is not, providing a stable flow of power. A new powerful battery would therefore spur the purchase of solar and wind turbines as a reliable alternative to electrical generation equipment powered by fossil fuel. For material scientists like Wyatt Metzger, a solar specialist at the National Renewable Energy Laboratory, batteries cannot improve quickly enough to help the market.[39]

At his lab, down the road from the Colorado School of Mines in the foothills of the Rocky Mountains, Metzger is working to revolutionize the solar panel. His technology relies on a mix of one of the world's rarest elements, tellurium, and toxic cadmium to revolutionize the solar energy sector.

Wyatt wants to transform the solar market, where 87 percent of all panels are made from high-purity silicon, and a dusting of a slew of minor metals, including molybdenum and niobium in the circuitry and indium on the panel. As MIT professor Robert Jaffe, co-chair of the 2011 American Physical Society report on energy-critical elements, notes, "Every watt of solar power requires minerals from peculiar parts of the periodic table." Wyatt and his team are developing an emerging technology called cadmium-tellurium thin films. The actual cadmium-tellurium layer is far thinner than a strand of hair.

Wyatt tells me they can be less expensive to produce than high-purity silicon. Other thin film technologies use gallium (the same materials used in LCD screens) and selenium, a sister element of tellurium. Their thinness and flexibility also open up new solar-powered options insofar as these panels can be rolled up, which makes them portable. This feature is key for military applications because getting fuel to conflict zones is expensive and dangerous.[40]

Although Wyatt speaks glowingly about the promise of thin film technologies, his current panels are less efficient than traditional silicon models, which means consumers would need more panels for the same power output—a costly proposition. In fact, many new technologies and especially emerging green technologies are not as efficient or cheap as current sources of energy. The cost per watt, the common metric for measuring solar panel efficiency, of cadmium-tellurium is $.65 per watt, and silicon is $.50 per watt. But this could soon change.[41]

Wyatt's U.S. government–funded team is making progress. Just five years ago cadmium-tellurium panels converted less than 10 percent of the sunlight they received into energy; now First Solar, a leading solar panel producer, converts close to 14 percent, and silicon panels convert 15 percent. The "levelized cost of electricity," a common metric for measuring efficiency across electricity sources, is 13 cents/kWh for solar approaching the 9.6 cents for coal.[42]

These metrics are important to Wyatt because his team wants cadmium-tellurium solar panels to produce energy as cheaply as coal. But some might worry that, when the total cost of solar power matches coal, solar technology will take off, thus causing a massive consumption of rare metals. As a society, we will replace coal mining with tellurium processing.

Environmentally, this is a great trade because it's better to process rare metals once and put them in a panel that makes energy continuously than to keep on extracting and burning new coal.

The solar panels use only a couple of grams of rare metals at most. And while the systems they are tied to may only use a handful as well, this adds up when millions of panels are needed. As the total amount of indium, selenium, and tellurium produced annually is about half the annual output of gold, the increased demand may cause instability. Relying on the rarest of the rare metals is not limited to solar power. One may not think much about the power consumption of an elevator, but in the buildings that have them, the elevator uses 5 percent of the structure's total energy use. Install a rare earth magnet motor in an elevator and it reduces energy use by half or more. This is a big saving, but it does not compare to the energy savings of new lights.[43]

Over the past two decades, lighting cost and quality have grown, and the average bulb is a lot greener. The new bulbs, compact fluorescent lightbulbs (CFLs) and light-emitting diodes (LEDs), are nearly 80 percent more efficient than older incandescent bulbs, which give off 90 percent of their energy not as light but as heat. Because lighting requires nearly one-fifth of the energy used in buildings in the United States, greener lights create real energy savings and reduce the need for forty new power plants. (To be sure they also make your wallet a bit lighter, so much so that when people move to a different house, they are less likely to leave these lights behind.)[44]

These lights are not your parent's bulbs; they're electronics—more akin to computers than what Thomas Edison invented over a century ago. In fact, LEDs have the same technology found in computers and at their core, they have gallium semiconductors. Lightbulbs have long been manufactured with some minor metals—from tungsten filaments in incandescent bulbs to various rare earth element powders in the fluorescent lights that jarringly light large box stores—but the new lights are softer on the eyes. This is because material scientists have learned how to deploy rare metals effectively, in particular by balancing the amount of rare earth element phosphors in them.[45]

Take GE's Reveal line of lighting. The light itself still produces a sharp, bright light, but neodymium in the glass provides a blue-tinted filter to absorb the yellow-green wavelength light, thus leaving a more pleasing light palette. Other rare earths like europium give a softer red and blue, and terbium provides green. The materials are so effective at filtering out certain strong lights that they have even been used in ski goggles.[46]

Legislation in the United States, such as the Energy Independence and Security Act of 2007 and similar laws in the European Union, mandate the use of these energy-saving bulbs. Paradoxically, such legislation has hastened the outsourcing of U.S. resource security insofar as these lights rely on foreign minor metals, which are increasingly made into products overseas. The world's largest lighting manufacturers, General Electric, Sylvania, Osram, and Cree, all moved facilities during the past seven years to the actual source of rare earths. While lower manufacturing costs are surely a factor in these moves, Sam Jaffe, a green research analyst at Navigant Consulting, says, "The primary reason many companies are in Asia has to do with material costs, not labor." And with Chinese export controls in place, Chinese domestic prices have been cheaper and supplies more reliable for green energy companies making their products there. With all these companies now manufacturing in China, the country consumes more than 80 percent of the world's rare earth phosphors for low-energy lighting systems.[47]

While the use of rare elements in lights brought about new technologies that use less electricity, reducing carbon dioxide emissions, sometimes just mixing rare metals with other metals can make a larger green difference.

Tadeu Carneiro, the CEO of the niobium giant Companhia Brasileira de Metalurgia e Mineração (CBMM), considers himself the head of a green tech company. "People say it's impossible to be sustainable if you have mining involved. Wrong. If I leave that niobium there, untouched in the sand," he says, gesturing to his mine, "the cars are going to be heavier, the structures in the buildings are going to be heavier, the pipelines are not going to be safe, and you [will] need more steel."[48]

And more steel means more CO_2.

If using less energy and therefore less CO_2 is a qualification for green technology, then the mining and processing of minor metals may be the one of greenest technologies of all. CBMM research shows that if new gas pipelines used 1 kilogram of niobium per ton of steel it would essentially reduce steel needs by half. Steelmaking is the most highly intensive CO_2-producing industry in the world, releasing 2.5 gigatons of the greenhouse gas annually.[49] Niobium not only reduces the overall amount of steel to be manufactured; its use in automobiles also helps to reduce the amount of CO_2 entering the air. A 10 percent reduction in car weight resulting from the use of niobium leads to a 6–7 percent increase in fuel efficiency. Niobium also finds its way into solar panels on the backside of the panel and has been shown in some labs to increase the efficiency of certain panels by nearly a third.[50]

No doubt a company's decision to use niobium and other steel strengtheners is not driven by green considerations alone, but by profit. Carneiro says this greening makes good business sense. A $9 investment in niobium, added to steel in cars, will reduce vehicle weight by 100 pounds. This will save one liter of fuel for every 200 kilometers driven. That saves 2.2 tons of CO_2 emissions over the life of the car, offsetting the amount of CO_2 necessary to produce the steel in the vehicle. The same is true for niobium in bridges, buildings, and pipelines.

Carneiro sees CBMM not only as a catalyst to a greener planet but also as part of a global system. "The challenge to be more efficient, to learn to do more with less is planetary. And niobium is there to help us in this transition."[51] And some places in the world, such as Beijing, are in desperate need of change.

For four days in Beijing in mid-February 2014, I didn't know whether the sky was clear or cloudy, the weather sunny or overcast. A thick fog cut visibility much the way a winter blizzard does, albeit without snowflakes. The air, with a faint hint of coal, left a yellow-beige film on cars, rooftops, and seemingly on me. Even a thin veil of smoke seeped into a hallway of the famed Lufthansa Center shopping mall.

China's demand for green energy is urgent. The life of its population depends on it. Air pollution in China led to 1.2 million premature deaths in 2010 and lung cancer rates that have doubled over the first decade of the millennium. And the air has gotten far worse since then. The patience of Beijingers to weather years of increasingly sickening air is limited and the government is working to change things.[52]

China targeted more than $294 billion in renewable investments between 2011 and 2015. Beijing's goal is to drastically raise the productivity of green strategic industries and it will invest $1.9 trillion in wind technologies alone over the next thirty-five years.[53]

This is a massive undertaking. China already leads the world in installed wind capacity, outpacing the United States by 50 percent. Such a drive not only could reduce China's growth of fossil fuel generation but also could make the country the testing grounds for green technology. Some companies in the West and in Japan are spending millions to wean themselves off rare metals in green products, but companies in China have no such reservations. The "innovation distortion" in the West and Japan leaves room for other companies, such as those in China, to develop a better product.[54]

Critics point out that resource shortages lead to price increases that spur more efficient use of resources and eventually new supplies. But as we have seen with catalytic converters and permanent magnet motors, some technologies rely on specific metals, and rare metal supply chains can't respond quickly to demand, especially for breakthroughs in technology.

The great paradox of green-tech breakthroughs might just be that we have the knowledge and the know-how, but we have not invested in the minor metal ingredients so that they are available at the right time, or even more important, at the right cost.

The biggest concern for rare metal supply lines may be that our new energy-saving gadgets work too well and that green tech will quickly become the best tech. Without foresight, we will be very short of the materials we need. It's a worry that military planners have, one that goes back to the 1950s when the United States was in a race to control the skies.

VIII

War Effort

Hard and Smart Metals

In 1959, Clarence "Kelly" Johnson, an aeronautical engineer at Lockheed Martin, designed the ultimate spy plane. The aircraft would fly more than twice as high as most planes of its day, at twenty-seven kilometers above the earth. And it would be faster too, traveling at more than three times the speed of sound. The plane would "dominate the sky for a decade or more" by evading Soviet missile systems while capturing images of the vast Russian land mass. Johnson's team at Lockheed paradoxically dubbed it the "Oxcart." There was only one problem: it couldn't be built.[1]

No material was durable or light enough to meet the plane's design specifications. In flight, the Oxcart's wings would approach temperatures of 260 degrees Celsius (500 degrees Fahrenheit) and the engine 650 degrees Celsius (1,200 degrees Fahrenheit). Aluminum was light enough, but at high temperatures it would warp because it loses strength at only 150 degrees Celsius (300 degrees Fahrenheit). The heat would then boil the fuel and turn lubricants to mush, while baking the pilot in the cockpit. Stainless steel could withstand the heat, but its weight would limit the plane's altitude, shorten flight distances, and require more fuel to keep it aloft.[2]

Two years earlier in 1957, William Senter, major general and chief of procurement and production for the Air Force's Air Materiel Command, was searching for a similar material—one that was practically weightless, incredibly strong, infinitely heat resistant, and able to be cast easily at negligible cost. Senter said that although he didn't have the material, he had a name for it: "Unobtanium."[3]

But by the late 1950s, metallurgists already knew about one possible material for the Oxcart; it was an obscure, unwieldy minor metal called titanium. Titanium is twice as strong as aluminum and matches steel's toughness at 40 percent less weight, and it doesn't corrode. More important, it is resistant to the plane's heat, melting at 1,660 degrees Celsius (3,020 degrees Fahrenheit), and is flexible enough to allow the plane to literally expand at extreme speeds and temperatures.[4] But titanium was no panacea.

Although titanium is the ninth-most-abundant element, it was a challenge to coax the metal from the mineral so the metal itself was rare. The United States had no reliable source of the metal despite government assistance to the industry of around $250 million, including more than $50 million to support 666 processing and mine development research activities in the 1950s. Those projects had yet to pay sufficient dividends to support the plane designs. The titanium then rolling off the line at Titanium Metal Corporation, the only U.S. supplier of the day, was of inconsistent quality. Low-quality titanium is extremely brittle and cracks easily, especially when dropped. So although the properties of the metal were ideal for flight, the material that the United States produced wouldn't entirely work.[5]

Because of these challenges, experts at the time optimistically predicted that titanium alloys would, at most, make up 30–50 percent of a plane's weight. The plans for the Oxcart were more ambitious: nearly the entire plane would be made from titanium. But without the high-quality rare metal, the plans for the plane would remain just that.[6]

Throughout history, a country's ability to harvest the power of the periodic table has translated directly to the success of its military, providing as much advantage in battle as do tactical decisions. Historically, civilizations that have mastered the art of making weapons from the most advanced metallurgy and materials of their day have dominated the societies around them.

Around 1400 B.C., 1,600 years after the Bronze Age began, metalworkers in a civilization located in present-day north central Turkey made an incredible discovery. The Hittites learned that when they heated and reheated iron ore over a hot charcoal fire, they could create an extremely strong metal. For centuries, bronze had been the mainstay of military weapons. But producing bronze hinged on finding deposits of copper and tin, which was difficult because roughly only 0.005 percent of the earth's crust is copper and 0.0005 percent is tin. Iron alternatively makes up about 5 percent of the earth's crust.[7]

Iron allowed the Hittites to produce weapons inexpensively, to equip legions of farmers and turn them into warriors. Hittite metallurgy unlocked a new military resource that quickly led to military success and eventually democratized war. As the Hittite blacksmiths migrated throughout the region, their metallurgical skill spread the Iron Age to the neighboring Assyrians and along with it, military superiority. The

Assyrians fashioned iron into other weapons, such as batter-
ing rams and shields. Their military advancements propelled
their conquests from the Middle East down to the Persian Gulf
and into Egypt as they overpowered the Arabs who were armed
only with bows and arrows.[8]

Centuries later, the Romans, who lacked rich ores like tin
and gold, established unmatched mining techniques, such as
hydraulic mining. They expanded their empire to bring rich
minerals back to Rome, which they turned into armaments
that they then used in battle.

Finding the right rare metal for the Oxcart was just another in a millennia-long line of metallurgical advancements
that have been key for military superiority. But these scientific
advancements did not stay confined to the military for long.
They became the basis for many technologies that underpin
our high-tech and green lifestyle. After more than a decade of
conflict and increasing military expenditures, we are once
again on the cusp of numerous material science breakthroughs.
These will filter down from the military not only allowing
technological advancements but also creating greater need for
minor metals.

But back in the 1950s, Kelly Johnson's immediate concern
was something less grandiose than the proliferation of the military's scientific advancements through society; he needed to
find a reliable supplier of titanium. In the name of national security, the CIA (Central Intelligence Agency), which worked
directly with Lockheed Martin, scoured the world to find a
source, and found only one: the Soviet Union. The agency reportedly used a network of shell companies and third-party
arrangements to purchase the titanium; in essence, it created
a spy program to buy material from the Soviet Union that
would eventually be used to spy on it.[9]

The next obstacle for the Oxcart was more formidable and has vexed every material scientist and engineer asked to use a new rare metal to replace another material, like Japanese corporations in 2010 who were trying to find a replacement for rare earth magnets. Using a new material in the plane's manufacture necessitated changing the plane's designs and specifications. The use of titanium "practically spawned its own industrial base," said Richard Bissel, the CIA's project manager. Approximately 2,400 people, from metalworkers to machinists, had to develop their own specialties to work with the new material. The use of titanium also meant that an assembly line wouldn't work; each plane had to be built by hand.[10]

Titanium is not an easy material to handle and most engineers and mechanics had never used it. Drill bits and tools kept snapping off because titanium is almost too strong. Cutting tools had to be sharpened every few minutes, hampering production. Even specially designed drill bits, which took months to develop, could drill only 120 holes before they too needed sharpening. The mechanics even needed new pens to draw on the metal because the chlorine in the ones they used created etch marks on the titanium. Cadmium-plated tools, including wrenches, had to be redesigned because bolt heads popped off the tools—microscopic chunks of cadmium would interact with the titanium bolts and cause them to break. Lockheed even used distilled water to wash the titanium panels because the local Burbank, California, water had too much chlorine, which caused the wing panels to warp like "potato chips" during heat tests. The Oxcart, eventually known as the A-12 and the "Titanium Goose," also needed different fixtures, lubricants, and fittings.[11]

The lessons learned from this use of titanium would soon filter down and transform the entire civil aviation world as it became part of airplane designs globally. This use helped turn titanium from a periodic table curiosity into a critical material. But at the time, the secrets of titanium helped the United States to own the skies in the early 1960s.

Roughly fifty years before the A-12, and nearly 3,500 years after the beginning of the Iron Age, designers at Krupp, a German weapons manufacturer, were struggling. They devised a massive 43-ton gun, called Big Bertha with tank-like wheels the height of a man. The gun was so large it needed close to 1,000 men just to set it up. But once in position Big Bertha could launch a 2,200 pound shell nine miles in a few seconds; the shell could blast through concrete fortifications. However, the gun could only shoot a few shots.[12]

The tremendous heat the weapon produced from firing its 42-centimeter-long shells melted the steel barrel. Engineers at Krupp discovered that spiking the steel with the minor metal molybdenum, with its high melting temperature of 2,617 degrees Celsius (4,753 degrees Fahrenheit), raised the melting point of the gun's barrel, effectively insulating it from the heat.[13]

The newly designed Big Bertha helped Germany to destroy Belgium's impregnable forts and then march into France in the early part of World War I.[14] Later, improvements in Big Bertha's design and production allowed the shelling of Paris in 1918, at the time shocking Parisians, who could not believe a weapon's shells could reach the city from over seventy miles away. But, like Kelly Johnson, the designers at Krupp were short on the rare metal they needed.

Before the war, Germany had trouble finding supplies of molybdenum. Few understood molybdenum's importance. But Germany knew its value and sought new sources. In 1915,

American Metals, a subsidiary of the German metal supplier Metallgesellschaft, discovered one source deep within Bartlett Mountain in Colorado, an old mine with limited molybdenum production. The German company bought it, planning to use American resources against the United States. But the plan never materialized, and Germany had to make do with the limited supplies it found elsewhere.[15]

Despite the lack of molybdenum and indeed many rare metals on its home soil, the German war machine was far better metallurgically prepared for battle than its neighbors. The well-trained German metallurgical engineers who were working in mines and managing them globally—even those in the British colony of Burma (now Myanmar), five thousand miles (eight thousand kilometers) away—supplied the minor metals that armed the homeland with the most advanced weaponry of the day.[16] So despite having a weak natural resource base, it had rare metals supply lines, superior metallurgical skills, and global investments to create an impressive fighting force.

This metallurgical savvy left other countries to play catch-up by reengineering German weapons to discover how minor metals were used in them. In addition, British officials could not understand why Germany was able to quickly replace the weaponry it lost. The British discovered that part of the secret lay in the use of tungsten ore from the heart of Great Britain. The British had practically given it away because they thought it was an impurity, failing to realize its importance.[17]

Due to its toughness, tungsten quickly became a military staple. It was so strong that its use in steel cutting tools reduced the time needed to shape a standard steel axle from 660 minutes in 1860 to 40 minutes by 1916. Rare metals like tungsten became so crucial to the war effort that the British outlawed their exportation to Germany. In the United States, authorities arrested three people for espionage because they had smuggled 200 pounds of tungsten headed for Germany.[18]

As the war continued, minor metals became increasingly more difficult for the Germans to import, so they became more efficient at processing these materials, and achieved the same results using less of the metal. At the same time, the molybdenum and tungsten-bearing metal alloys found their way into the weapons of Germany's adversaries. The United States, with its vast rare metal resources, had a supply advantage that would become crucial to ensuring a material advantage over Germany. At the end of the war, Colin Fink, president of the American Electrochemical Society, boasted, "It may some day be said that tungsten made democracy possible." But he failed to acknowledge that Germany's use of tungsten may have also made the war possible.[19]

The demand for minor metals had increased because of the war and the use of new metallurgical processes to employ them. For example, the use of flotation, a method for separating various metals from minerals, increased from 1 million tons to 150 million tons annually in just a few years. Scientists also developed new ways to form and molt molybdenum, which helped in developing new nonmilitary uses, first in cars of the early 1920s and then in other tools.[20]

And now, the world consumes more than 250,000 tons of molybdenum annually. Because of its widespread use—in everything from fertilizer nutrients and flame suppressants to lubricants—some do not even consider molybdenum a minor metal. It is now a critical element in high-strength steel and it forms the backbone of some of the world's tallest buildings such as the Petronas Twin Towers in Malaysia and the Jin Mao Tower in China. Similarly, Kelly Johnson's titanium is now at the heart of aerospace technology, and also appears in paints and medical implants.[21]

This metallurgical route from sword to ploughshare has existed for thousands of years—even the Hittite advent of low-cost iron did more than just lead to military advantage; it helped to put high-quality tools like saws, drills, and screws into the hands of its citizenry. As the Task Force on American Innovation notes, "civilian applications of technologies intended originally for military purposes have become staples of the nation's economy and modern life."[22] In the past minor metals such as titanium were important because they made weapons harder, stronger, and heat resistant. Now minor metals are making weapons smarter.

In 2013, I met Dominic Boyle, a minor metals trader. Though Boyle is British, he speaks fluent Japanese and acceptable Chinese—important languages to master for a trader who wants to keep good relations with his closest customers. Global conflict has proved to help sales. He has watched the price of germanium, a metal his company trades, quadruple over the past decade as the market doubled due to rising tensions in Asia and greater military spending and conflict in the Middle East. Speaking at the Shanghai Metal-Pages conference in 2013, he tells the audience that strong demand for germanium will continue. It coats fiber optics cables and high-speed circuitry, but its recent demand comes from the defense sector.[23]

Germanium is at the heart of thermal-imaging systems (think: night vision goggles) in aircraft, ships, and tanks as well as weapon sights mounted onto rifles that allow for more effective reconnaissance missions. It's a lustrous silvery-white metal but is transparent in infrared and helps translate infrared radiation into images.[24] Its properties seem tailor-made for many military applications.

U.S. military demand for these applications soared after the invasions of Iraq and Afghanistan, and with it, the sale of germanium. The overall U.S. demand for germanium in thermal optics used in defense applications jumped from 5,000 tons in 2003 to about 30,000 tons four years later. Approximately half of all germanium consumed in the United States, and one-third globally, goes to thermal imaging. In 2009, the U.S. Defense Department spent almost $1 billion on night vision goggles and thermal weapon sights.[25] War has transformed the germanium market. And this is not the first time.

Nearly three-quarters of a century earlier, in 1942, scientists at the Manhattan Project in Los Alamos, New Mexico, were mixing isotopes of the rare metal uranium to create the atom bomb. At the same time, another set of military-supported researchers were conducting equally important research at Purdue University. Their goal was to improve the performance of radar.[26]

They tested the properties of germanium, an obscure minor metal at the time, which had little usage up to that point. The diodes in use until then burned out too quickly, especially with the emerging radar systems that used shorter wave lengths to produce enhanced detail. The researchers were looking to create a powerful device that would allow charges to flow only in one direction, the way a heart valve allows blood to flow out but not back in. But what they discovered would ultimately usher in new electronics and the Rare Metal Age.[27]

Unlike copper and aluminum, which allow electrical charges to quickly pass through them in either direction, or plastic and glass, which don't conduct charges at all, germanium is a semiconductor. The Purdue researchers discovered that if they introduced slight impurities such as arsenic and phosphorus into high-purity germanium, they could create a diode that would allow charges to flow only in one direction, which made it possible to convert alternating current to direct current and radio signals into audible sounds.[28]

This seemingly insignificant finding produced a diode that was ten times more resistant to burnout than previous ones. It laid the foundation for transistors and integrated circuits—the semiconductors critical to the electronics we use now. But back in 1942, the scientists' goals were far more modest; they just wanted to see the skies over Germany. They had no idea that they would change the way we live and also drive up the demand for germanium more than twentyfold in less than a decade.[29]

Although the U.S. military is reducing its use of germanium in thermal imaging equipment as its wars end, a potential conflict is spurring new demand. Rising tensions between China and its neighbors, most notably Japan, over territorial ambitions in the South China Sea, are currently leading the demand for germanium. In early 2014, a Chinese-based infrared supplier commented that infrared orders from the national defense sector had increased markedly over the year. Military budgets in the region have in some cases doubled or nearly tripled over the past ten years, and in 2014 China increased spending by another 12.2 percent. Rising tensions and budgets mean good business for metal traders like Boyle. Even small conflicts can be an opportunity for these suppliers.[30]

"When the war broke out between Israel and Lebanon [in 2006], phones lit up," says Ed Richardson, vice president of the rare earth magnet producer Thomas and Skinner. "They [producers] pulled in basically a year's worth of consumption into about a month, because they needed magnets for their systems to work."[31] The challenge now is that so much of our military equipment also relies on a multitude of resources that may become thin during times of conflict when demand for them is at its highest. All this reliance on materials means that some of the most important facilities in modern warfare are now very far from the front lines.

Today's most important battlefields are not in the places where missiles land or guns fire, instead they are in the material science labs of places like Sandia National Laboratory in California, Beijing's Tsinghua University, and the United Kingdom's BAE Systems Lab, where researchers are racing to develop materials that will propel their country's military ahead in the next generation of warfare.

Robert Latiff, a retired major general says, "It was the advent of advanced electronics that changed the nature of wars." Latiff, who holds a PhD in materials science in addition to two stars, believes that understanding the power of rare metals helped the United States switch from the hydraulic and mechanical weapons of World War II to today's electronics-based weaponry that uses actuators and sensors, all of which require minor metals. This type of warfare, Latiff notes, relies on electronic hardware for intelligence gathering and surveillance. This means that we need not only strong minor metals but also metals that help to regulate the flow of electric current efficiently.[32]

"Without some of these minor metals you would have to go back to 1960s or 1970s performance," Latiff tells me. Nearly all of our systems today depend on rare metals. "It is inconceivable we could achieve [what we have as a military] without having the material science capabilities that we do."[33]

The strongest militaries of today are the ones that can harvest almost the entirety of the periodic table. A mixture of cadmium and tellurium is at the heart of radiation-detection systems as well as in baggage-scanning and dirty-bomb-detection machines. Missile guidance targeting and control systems contain a slew of rare earth metals, including terbium, yttrium, and europium. And tungsten is crucial for armor-piercing bullets and drones that shoot down GBU-44 Viper Strike missiles.[34]

Although these new materials have enormous performance benefits, they create a challenge for the U.S. military—a reliance on minor metals not found or produced in the United States. The United States must import at least 75 percent of over twenty-five different rare metals; this is up from fifteen metals in 1995.[35] Similar to Germany's dependence on imported molybdenum and tungsten during World War II, a growing reliance on rare metals poses a threat: the potential for a minor metals shortage at a time when they are needed the most for defense.

"This near-total dependence for critical components and raw materials creates worrisome risks," notes former brigadier general John Adams and Scott Paul in *Politico*, "Our security and our ability to develop future battlefield capabilities are dependent on potentially unreliable supplier nations who might not have our best interests at heart in a crisis situation."[36]

The military's largest systems are often the most vulnerable to minor metal shortages because of their vast and complex needs. According to the U.S. government, just one of the nearly fifty planned nuclear-powered submarines uses approximately

9,200 pounds of rare earth materials; every one of the seventy-seven DDG 51 Aegis destroyers uses 5,200 pounds, and each of the forthcoming F-35 Lightning II aircraft requires approximately 920 pounds, which enables the pilot to do nearly everything from starting the engines to controlling wing flaps for landing to assisting the electrical interface between the plane's controls and its components.[37]

It is not just rare earth elements. The newest weapon systems like the F-35 are flying periodic tables. The plane is almost a quarter titanium, which keeps the frame light and heat-resistant. The nuts and bolts that hold the plane together are made from beryllium because of the metal's strength and heat resistance; gallium helps to amplify the radar's signal; and lithium provides a high energy density battery. Tantalum, because of the material's ability to hold a charge, sits inside the capacitors needed for laser targeting, controls, and cockpit displays.[38]

To reduce its risk of relying on precarious supply lines, the U.S. military is assessing its resource demands and the reliability of its minor metals suppliers. But it's a daunting task to ascertain every element and alloy in every component in every system as well as where it comes from. The military, a high-tech workforce of over 1.4 million people, uses nearly all the metals that are commercially available in its weapon and computer systems.[39]

The computer chip manufacturer Intel needed more than two years to understand its supply line well enough to ensure that no tantalum from the Congo was in its microprocessor products, and it was the first in its industry to do so. But unlike Intel, which examined only its microprocessor supply line, the U.S. military has millions of components to assess. It will take far longer and the process may be endless because the military's hardware needs are constantly evolving.[40]

Scientists are continuing to find new ways to save on costs and improve the efficiencies of existing hardware by using more and different minor metals. For example, titanium replaced aluminum in the hatch on the M2 Bradley Fighting Vehicle, which created a 35 percent savings in weight and protected those inside the M2 more effectively.[41]

To reduce resource supply risk, the military is trying to switch to more commonly used, domestically produced resources. But this too takes time. The U.S. Government Accountability Office estimated in 2010 that it would take at least fifteen years for the Defense Department to revamp its supply chain so it was not reliant on foreign supplies of rare earth magnets. However, General Latiff commented to me, "I'm not sure we'll ever completely get away from foreign sources." So, in the meantime, Chinese rare earth magnets are critical components of the most advanced U.S. weapon systems, including the F-35.[42]

With the ever-increasing demands of the military and the finite list of ingredients, the replacement options are few. It's not possible to simply replace one commodity for another in every application. Thomas Graedel, at the Yale School of Forestry and Environmental Studies, found that out of sixty-two metals he studied, not one metal has a substitute that can replace it for all its original major uses. What's more, twelve have no suitable substitute for any of their properties as shown in Figure 3. If the food and beverage industry still can't effectively replace sugar or the transportation industry, oil, it would be unreasonable to expect the military to replace its use of rare metals.

"Going from Mach 2 to Mach 4 or 5 presents an entire new set of [material] performance specs," a material scientist at the

H								
Li 41	Be 63							
Na	Mg 94							
K	Ca	Sc 65	Ti 63	V 63	Cr 76	Mn 98	Fe 57	Co 54
Rb	Sr 78	Y 95	Zr 66	Nb 42	Mo 70	Tc	Ru 63	Rh 96
Cs	Ba 63	•	Hf 38	Ta 41	W 53	Re 90	Os 38	Ir 69
Fr	Ra	••	Rf	Db	Sg	Bh	Hs	Mt

* Lanthanides	La 75	Ce 60	Pr 41	Nd 41	Pm	Sm 38
** Actinides	Ac	Th 35	Pa	U 63	Np	Pu

Substitute Performance

Excellent |__|__|__|__|__|__|__|__|__|__| Poor

0 10 20 30 40 50 60 70 80 90 100

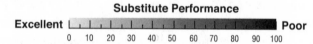

Figure 3. The periodic table of the substitutability of various metals on a scale of 0 to 100.

Source: T. E. Graedel, E. M. Harper, N. T. Nassar, and B. K. Reck, 2013. "On the Materials Basis of Modern Society," in *Proceedings of the National Academy of Sciences.* doi:10.1073/pnas.1312752110.

								He
			B 41	C	N	O	F	Ne
			Al 44	Si	P	S	Cl	Ar
Ni 62	Cu 70	Zn 38	Ga 38	Ge 44	As 38	Se 47	Br	Kr
Pd 39	Ag 44	Cd 38	In 60	Sn 36	Sb 57	Te 38	I	Xe
Pt 66	Au 40	Hg 45	Tl 100	Pb 100	Bi 46	Po	At	Rn
Ds	Rg	Cn	Uut	Fl	Uup	Lv	Uuh	Uuo

Eu 100	Gd 63	Tb 63	Dy 100	Ho 63	Er 63	Tm 88	Yb 88	Lu 63
Am	Cm	Bk	Cf	Es	Fm	Md	No	Lr

defense contractor Raytheon told me. Increasingly stringent specifications mean that the military will need more, not fewer, minor metals in refined, pure amounts. This poses a challenge because only a few companies in the world can refine or produce the high-quality minor metal material to meet this stipulation. As Raytheon notes in its company magazine, "The demand for stronger, lighter, greener yet less expensive materials continues to outstrip availability." Despite the pressure to use new materials, the military paradoxically prefers proven material from reliable suppliers. "The military doesn't like cutting edge," says Yossi Sheffi, director of the MIT Center for Transportation and Logistics.[43]

"If you have a new material you want to get into a weapon system, it takes a long time, twenty years on some occasions," says General Latiff. The reason is simple: if the military is going to invest billions in a new plane, it doesn't want to rely on unproven materials. Since the military is unable to switch materials quickly, it's vulnerable to resource shortfalls. But something else could be more worrisome for the U.S. military than a shortfall of materials: the country that produced the most efficient war fighting force on the planet with unmanned flight, precision guidance missiles, and radar may be losing its competitive edge.

The limits of creating weapon systems are not our imaginations but our engineering ability, and that starts with material scientists themselves. But a lack of metallurgists and material scientists graduating from U.S. universities, like Caelen Anderson at the Colorado School of Mines, whom we met in Chapter 4, may be one of the U.S. military's greatest weaknesses. As General Latiff says, "Military materials scientists are critical." He tells me that without them we are depending on other countries for our military development because we have to know the capabilities of the materials before we can put them into a weapon system. Expanding material science capabilities is crucial because armies cannot use twentieth-century materials to defeat twenty-first century threats. Increasing our material science knowledge is not just critical to defending against threats on the battlefield, it's central to addressing the most existential threat we face: climate change.[44]

IX

Sustainable Use

The Environmental Calculus of the Rare Metal Age

Zhang Yang'e, a Chinese farmer from the remote village of Dingnan in Jiangxi Province, lives some ten meters from a rare earth element mine. The operation poisoned her well and killed her crops. "The water used to taste sweet and our neighbors all loved it. But now it has been become undrinkable," she told a local reporter. "Even my vegetables withered after I watered them with well water," Zhang says.[1]

People no longer swim in the rivers around Jiangxi Province. They have turned toxic, killing most of what once lived in them. From above the rolling green hills of Jiangxi, a province roughly the size of Greece, you can see the devastation of the mines. Some leave bare swaths of beige clay, as if the miners had peeled off the skin of the earth. Other mines slice around the contours of the hills in unnatural but elegant tiers, like smaller curved versions of the CBMM (Companhia Brasileira de Metalurgia e Mineração) mine in Araxá. Each mine follows its own unique path according to the land's geology, but the mining process is the same—and so are the scars.

Even though you may never have been to the hills of Jiangxi your footprint is there. Nearly all of your electronics contain specks of metals from those mines because, as we saw in chapter 4, the hills in southern China produce nearly all the world's dysprosium needed in all our high-tech goods. When we import Kindles, we export pollution, including wastewater, carbon dioxide emissions, and acid mine drainage.[2]

To gain access to rare earths, Zhang's neighbors dig eight-foot-deep holes in the hillside and pour down ammonia sulfate acid into them. The hills are made up not so much of rock or dirt—but a rather brittle sandy clay that the acid easily penetrates. The rare earth elements, such as dysprosium and terbium, in Jiangxi are easy to access because of the weak chemical bonds between the rare earth minerals and the clay. So the mining process is similar to pouring water on a mound of potter's clay and then watching the pile disintegrate.[3]

Once the acid has washed away most of the clay, miners trudge through a swampy mess of brown muck that compresses with each step. Their clothes splattered beige from the wet clay make them look more like painters, as they hoist white bags of brown sludge over their shoulders. They lug them to a network of tropical aqua-blue and brown treatment pools of solvent connected to one another by white PVC piping. Its deep rich blue liquid is oddly alluring, like a small child's water park. But this jerry-rigged web of plastic on the slopes of the hills is no playground. It chemically helps separate the rare earth elements from other minerals in the rest of the clay. The material that's left after the acid bath is baked in a kiln to form a concentrated mix of rare earths. It's a far easier and cheaper process, even compared to CBMM's in Brazil.

But the amount of waste from the operation is staggering. Only 0.2 percent of the mined clay contains rare earth elements. This means that 99.8 percent is discarded waste called "tailings" that are dumped back into the hills and streams. They seep into streams and wildlife habitats, bringing some of the chemicals Zhang complains about with them. Because the mountains are gouged of large chunks and replaced by weak processed clay, they are prone to landslides. According to Beijing officials, 2,000 kilograms of tailings are created to produce every kilogram of rare earths. The process also leaves a mix of ammonium and heavy metal among other pollutants in the sediment runoff that affects Zhang.[4]

Paradoxically, the earth that is put back has a greater volume of clay than what the miners originally removed. Air seeps into soil when it is no longer compacted, causing it to swell. This swelling is why there always seems to be more dirt and rocks leftover after you fill a hole you have just dug. It is also why miners cannot put all the soil back on the mountainsides from where it was taken.

Chinese officials estimate that it would cost about $6 billion to clean up the polluted area around Jiangxi, but the figures on regional damage are speculative and may understate the problem.[5] With the Jiangxi rare earth mine critical to our high-tech lifestyles and the region, Zhang's concerns are unlikely to be addressed quickly.

For a glimpse at what the future could hold for Zhang and her neighbors in Dingnan, look no further than 1,400 miles north to Dalahai, a village on the outskirts of a tailings dam from Bayan Obo mine. Li Guirong, age sixty-five, has watched the region change over decades as this mine, near the border of Mongolia, has become the largest rare earth mine in the world in addition to supplying iron ore, at the expense of those living around it.[6]

"We knew something was wrong sometime in the late 1980s when the trees and local vegetation blossomed but didn't bear any fruit. Later, they stopped growing at all," Li tells local media. Li's neighbors developed skin rashes, respiratory diseases, osteoporosis, and cancer. Villagers had trouble keeping their teeth.[7]

The ore is richer in the north, although it's rockier, so it needs to be crushed and then treated with sulfuric acid to help crack bonds that bind rare earths to the minerals. The more time-intensive process also releases radioactive thorium that exists in the local rocks and in nearly all other rare earth deposits. To produce one ton of rare earths in Bayan Obo, workers use more than four tons of sulfuric acid and one ton of hydrochloric acid and sodium hydroxide. The chemicals create a difficult work environment. "The pungent smell of the powder makes me cough from time to time, and the thick mask does not work at all," a worker at the site told local media. The Chinese Society of Rare Earths estimates that the process produces 75,000 liters (nearly 20,000 gallons) of acidic wastewater and one ton of radioactive residue per ton of rare earths.[8]

In Baotou, the home of the Bayan Obo mine, the separating and smelting processes generate wastewater containing high concentrations of ammonium nitrogen and radioactive residues. These are dumped into a tailings lake, leaving the air so acrid that it burns the eyes. Because environmental precautions have not been high on the mine operator's agenda, the tailings lake wasn't properly lined; its contents contaminated the surrounding soil, groundwater, and vegetation, turning Dalahai into one of about 450 "cancer villages" in the country.[9]

It's impossible to produce minor metals without some environmental effect. (The same could also be said for modern farming or manufacturing.) But the process does not have to be as environmentally damaging as it is in China. When countries choose not to produce their own rare metals in environmentally conscious ways, they are essentially outsourcing it to places like China, which leads to greater global environmental damage. As we will see, our current system of rare metal production and use is focused on the short term and is wasteful. If this continues apace, our demands will cause more damage to our environment than necessary. And despite minor metals' significant effects on the environment, we are very poor at recycling them. But recycling alone is not a panacea. It too has its own environmental consequences and also cannot by itself supply all of the world's minor metal needs.

Minor metals are a slim portion of all metals produced, but they often have outsized environmental impact. They frequently require much more processing than base metals like copper and zinc, and they use greater amounts of solvents.[10] We must understand the environmental implications that come with mining more of these minor metals because the global environmental impacts of production cannot be a separated from the gadgets we buy, which, by extension, means we are all responsible for the environmental effects of mining these metals. Kohmei Halada knows this well.

Halada works at Japan's National Institute or Materials Science, with a James Bond–type sounding job title "senior scientist with special missions." But his job is more straightforward than his title. He tabulates the environmental impact of our lifestyle choices. He is part of a growing field of researchers who look at the life cycle of our products.

Every step in the life cycle of our gadgets—production, use, and disposal—produces greenhouse gases. But as consumers, we don't see the emissions. Unlike the short tailpipe on our cars, the long tailpipe of our high-tech lives obscures its exhaust. We might assume that the greatest amount of electricity used and its subsequent contribution to greenhouse gases occurs when a gadget is in our hands or in a wall charger. After all, it's the only use of electricity we see. But Nokia and Apple found that a mere 15 percent of the greenhouse gases generated by the entire life cycle of many of their products come from the electricity needed to charge them. This means that roughly 85 percent comes from their manufacturing, shipping, and disposal. This is why it is often greener to use an older product than to buy a newer one, despite the apparent energy savings of improved batteries and shorter, less frequent charges.[11] To understand a product's true environmental impact we must also consider the pollution generated from manufacturing and disposal.

Halada knows from research what Zhang, the villager in Jiangxi Province, has seen firsthand: producers make a lot of waste to obtain a small amount of metal. Take a product that many of us would consider low-tech: a 3-gram platinum wedding band. To arrive at what Halada calls the total material requirement for the ring, he quantifies the amounts of all resources used: the soil consumed and moved to access the metals, the amount of coal burned to create energy to process the material, and the water used in the production process. By this arithmetic, one 3-gram ring requires 3.6 tons of material. In contrast, the total material required to produce 1 ton of iron is about 8 tons. But even platinum's resource demands pale in comparison to other metals. To make 1 ton of germanium requires 120,000 tons of material. (Since germanium is a by-product, some of that other material includes other metals.)[12]

According to Halada, if you tally up the entire environmental influence of producing a kilogram of rare earth elements, it is roughly equivalent to the effect of producing 1 ton of iron. Despite such precise figures, Halada and fellow researchers concede that comparing the effects of different metals is subjective because some metals have low carbon footprints and higher localized effects, such as wastewater runoff or deforestation. However, assigning an amount of material needed to make a certain quantity of a metal provides perspective. On average, industrial products use a total of materials thirty times their own weight.[13] But for green and high-tech goods such as phones, LCD displays, and their internal components, the factor can increase into the hundreds and more because high-tech products use many rare metals. A basic mobile phone weighs 56 grams, but it takes 31 kilograms of resources to make.[14]

Rare metal production, like most mining, can lead to acid runoff, metal contamination, increased air emissions, and in some cases radioactive waste. Every step of production has an environmental influence—from the clearing of land to the heavy use of water in mining. Even the mines themselves, which are often in remote locations, need their own infrastructure—from roads to sewer systems.

Ronald R. Cohen, professor at the Colorado School of Mines, explains that just digging earth, the first step in mining, exposes sulfide minerals to air and water. This exposure can lead to "acid rock drainage"—which some call acid mine drainage, as the mining speeds up otherwise relatively benign

natural processes that usually occur over millennia as more
surface of rock comes in contact with water. To demonstrate
to his students how this occurs, he puts a stick of chalk in water
and asks students to estimate the amount of the chalk's surface area that interacts with the water—a couple of inches. Cohen then breaks up the chalk, and tells the students that the
chalk's surface in contact with the water increases by several
times. But if he were to grind up the chalk instead, he explains,
the surface area in contact with water would increase millions
of times. It is these refined minerals and their increased contact with the water that lead to the acid drainage. The water
turns the sulfides into sulfuric acid and it poisons local waters
for tens to hundreds of years or more as new water continues
to come in contact with the acidic rock.[15]

Cohen tells me that ten thousand to twenty thousand
streams in the United States are now lifeless because companies failed to take precautions to prevent or remediate acid
mine drainage. Once the acid hits the stream, the water
cannot be used for agriculture, wildlife, or even for industrial
purposes because it can ruin equipment. Nearly a century
ago, when people in the industry discovered the acid problem,
they addressed it by fixing the equipment. They developed new
acid-resistant pumps that the water couldn't corrode. Reducing the acid was an enormous task, so they continued to pump
it into local streams. Unfortunately, the damage from acid
drainage can last for centuries. Previous poor mining practices and weak oversight took an environmental toll especially
as inadequate tailing ponds allowed hazardous materials to
leak off mine sites and mining companies left mine sites in a
lifeless state without reclamation.[16] Metal processing also had
profound environmental effects.

In the early 1900s, Osaka, Japan, became widely known as "Smoke Capital" due in large part to cumulative emissions from the metal industry processes. Dokai Bay, outside the country's southern Japanese city of Kitakyushi, was named "The Sea of Death" because of the beige and green mix of chemicals and waste lurking in the waters from the iron, steel, and electrical industries.[17] Lax regulation not only affected the environment; it caused tremendous pain to its people. During the 1930s, cadmium from a lead-zinc smelter leaked into Toyama Prefecture's water supply, and from there, leached into the region's rice fields. Eating cadmium-enriched rice caused several hundred local women severe pain as the toxin seeped into their bones and softened them, and left some women permanently hunched over at the lower back. The pain was so excruciating that it became known as "itai-itai," meaning "ouch-ouch."[18] Despite the horrific nature of the incident and a polluted environmental legacy, it was still years before Japan became greener, in part by importing more metals processed elsewhere and by imposing stricter environmental regulations on its mines.

The transition to cleaner metal production methods was a global movement that actually started elsewhere with the rise of environmental consciousness in the late 1960s and 1970s in the West. A flurry of environmental legislation, especially in the United States, compelled many companies to begin to minimize their environmental effects. Companies began adhering to a core of international environmental standards. The push for cleaner mining was not just because of ethics, it was also due to economics. As water, energy, and fuel resources have become more costly, efficiency has not just become good for the environment, but good for business.

But at the same time that mining companies are becoming more environmentally conscious, mining is becoming more environmentally challenging. Newer mine sites are often more complex because deposits are deeper, more dispersed, or have lower ore grades with smaller grain sizes, which leads to more resource-intensive processing methods. To produce the same amount of minor metals that we use today, we will burn more carbon dioxide to supply power for mining operations and we will create more waste—chemicals and tailings—tomorrow.

In many cases, miners are creating far more waste because they have to dig deeper. As the percentage of metal in the ore drops, companies need to process more ore to make the same amount of metal. One study notes that as copper ore grades fall from 3 percent to 0.5 percent, the energy needed to process it increases more than fivefold.[19]

The real challenge for many mining companies now is not just to handle the environmental repercussions of current operations, but to overcome the sector's previous environmental negligence and lack of openness. The metal businesses' past transgressions often thwart today's minor metal development.

In 1982, on the fringes of western Malaysia's jungles, Mitsubishi Corporation began a joint venture processing monazite, a mineral containing rare earth elements but also high levels of naturally occurring radioactive elements. Locals complained that the company initially dumped waste in plastic bags in an open lot and encouraged them to use the material as fertilizer. In 1985, the company reportedly left radioactive waste in drums in the open outside the plant.[20]

Local residents of the adjacent town of Bukit Merah blamed the refinery for a spike in radioactivity and related rare diseases. The eleven thousand residents claimed that Mitsubishi was to blame for a local increase in rare diseases including pineal gland tumors and eight leukemia cases over five years where previously there had been none. Mitsubishi closed the plant after legal battles in 1992, and spent $100 million to clean up the area, which meant digging up radioactive soil twenty-five feet below the surface, down to the bedrock. The firm contributed more than $150,000 to local schools but denied responsibility for the pollution.[21]

The plant didn't just leave a radioactive legacy; it left a toxic environment for those companies that followed. Mitsubishi's reputation dogged the Lynas Corporation when it tried to open a rare earth processing facility in the country in 2011 to process imported ores from Australia. Locals felt they were being treated as a dumping ground for Australia's radioactive waste, as they once had been for Japan's. In fact, Australia had already rejected a rare earth processing plant in the 1980s because of environmental and health concerns related to processing ores, similar to those in Bukit Merah.[22] It was likely that the challenge of opening facilities in Australia and Japan helped send Lynas to Malaysia.

Despite assurances from the Malaysian government that Lynas could begin processing, environmental activism almost doomed the project, even with Lynas's extra safety precautions. The company found it increasingly difficult to obtain operating licenses from a government besieged by environmental protesters, even though the minerals heading to the new plant produced about thirty times less radioactive material than those that Mitsubishi had processed.[23]

To avoid similar opposition to new projects, companies are spending millions of dollars annually on complying with environmental regulations. The rare earth producer Molycorp, based in the California desert, spent over $1.5 billion to develop a state-of-the-art system to conserve water and recycle waste. But no system is perfect. Molycorp was still cited in 2014 for inadequate management of hazardous waste. Even Teck Resources, which received a Prospectors and Developers Association of Canada environmental award, recently revealed that its subsidiary had polluted for more than a century by discharging solid and liquid waste into the Columbia River.[24]

It is in a company's interest to reduce the amount of water it uses and reuse acid rather than dumping it into the environment. But of course, it costs companies money to be efficient: the equipment they need to recycle high-quality acids and water is expensive. These additional expenses proved to be a hurdle for Molycorp and other rare earth producers outside China during the 1990s. The Chinese didn't buy expensive processing equipment because they didn't need to adhere to stringent environmental regulations that others did.[25]

But Cohen from the Colorado School of Mines believes that producing metals can be done responsibly, if only consumers would be willing to pick up some of the cost. If the economic benefits of environmental compliance are not clear, the only things that keep mining operations clean are the morals of the company's management or the enforcement of environmental regulations. And in some places both of these are scarce resources.

Instead of mining new materials, many hope that recycling our products will be our environment's savior. Metals in essence are lent to our products, rarely are they consumed like oil or coal. A two-year-old iPhone contains slightly more metal than a brand new iPhone, which over time has become slightly more metal efficient to save money. The task is to find a way to remove metals from the old phones so that they can be used again in the next product. While recycling metals will reduce the environmental influence of mining by necessitating less of it, recycling also requires energy and resources. It is often better than extracting new metals, but it is not without repercussions.

"I love scrap," David Gussack tells me over beer at a bar near Grand Central Terminal in New York at the Minor Metals Trade Association's annual happy hour. David, whose business card calls him a "metallurgical specialist," is a recycler who buys scrap metal from manufacturing companies that make everything from tantalum to niobium. He reprocesses it into sellable material at his 35,000-square-foot facility, just off the Florida Turnpike in Pompano Beach, often selling it back to the companies he bought it from.

Gussack asks if I know what sputtering is, then explains a process for coating an object with metal. Sputtering is messy. Think of a Jackson Pollock painting, spattered with various colors. That process might create a work of art but it also makes a mess, leaving behind a lot of paint on the walls and floor. Gussack's job is the equivalent of coming into Pollack's studio, ripping up the floorboards and taking down the walls, and scraping off the paint for recycling. He profits from this cleanup. Gussack takes the refuse, which could be shavings of steel or frothy liquids, and turns it into usable metals—and tens of millions of dollars annually for his company. His work also provides a service to the environment, by reducing the need to mine more ore. Gussack's business has become so profitable to manufacturers that it blurs the customer/client relationship. They are symbiotic partners, buying and selling from one another.

Because rare metal waste now has increasing value, manufacturers' thinking about using material more efficiently has changed. When the platinum refining facility of U.K.-based Johnson Matthey instituted a policy mandating that employees wipe off their feet before leaving the facility, they collected enough dust each year to buy a well-appointed Volkswagen.[26]

Although this industrial recycling is a positive environmental development, it also highlights the inherent inefficiencies of the manufacturing process. Just as we learned in chapter 4 that material is lost during the metalmaking process, the same is true in the manufacturing process. Take rare earth element magnets for example. Manufacturers lose about 15–50 percent of their materials as they carve and shape them to the correct size. Steve Constantinides of Arnold Magnetic Technologies tells me that the percentage of waste has only marginally improved since he started working in the industry more than two decades ago.[27] Whereas companies have become better at wasting less in the manufacturing process, the smaller sizes and exacting shapes that companies now demand continue to challenge magnet makers, which leads to more waste.

While Gussack harvests his resources from factories, 7,500 miles (12,000 kilometers) away from Florida, Hitachi works on another challenging effort that will be key in transforming our resource use. The company's eco-recycling facility sits on the edge of Tokyo Bay on essentially recycled ground that was part of the bay just a generation ago. Their work—"post-consumer recycling"—is what transforms the plastics, air conditioners, and computers that we abandon. Eco-recycling, of course, cannot pick and choose what is recycled; its task is far more complex and less efficient than Gussack's industrial recycling.

When I visited the Hitachi facility in 2011, the director said that it recycled more than 95 percent of everything it received—although his definition of recycling includes turning a good portion of that material into pellets burned to make electricity. Tokyo Foundation's researcher Hikaru Hiranuma calls recycling "urban mining." Hiranuma estimates that Japan has more metals like silver, gold, and indium sitting in electronic gadgetry around the country than in the reserves at many mine sites globally. He believes that harvesting these resources is crucial to Japan's future; it will make the country greener and more resource secure.[28] But turning mobile phones and LCD screens back into metals is far more complex than doing a simple syntax change.

Just getting the raw material of urban mining—the iPhones, air conditioners, and electric toothbrushes—to recyclers like Hitachi is a challenge. Even in Japan, where almost everyone is recycling-conscious—people even remove labels from plastic bottles before throwing them in recycling bins—the country only recycles 10 percent of personal computers.[29] Wasting resources is, of course, not just a Japanese concern. It is a global theme. According to the U.S. Environmental Protection Agency, the United States discarded 75 percent of 2.4 million tons of electronics in 2009, the year after the iPhone came out, without any form of recycling. Often, the first obstacle to recycling involves getting the product to the recycler.[30]

Ken Deckinger is like most Americans, except that he started several successful Internet companies. When the tech entrepreneur went to upgrade his iPhone, he had no idea his phone carrier would give him more than $200 for his old one.

Previously, Deckinger just threw them in the trash along with most of his other electronic equipment, save an old MacBook and even older ThinkPad that he meant to donate. The truth is that even the most savvy tech folks do not reuse or recycle outdated electronics because they do not know where or how to do so.

In 2014, a survey by EcoATM showed that only 22 percent of Americans surveyed recycled their old mobile phones or tablets. Even though most Americans believe that recycling is good for the environment, less than half would consider recycling these gadgets. Other surveys show that half of respondents did not know where to recycle e-waste, didn't think about it, or didn't know it was possible. Most keep their old phones as spares or give them away. People complain of the hassle to take it to a collection site—if they even know where one is. This lack of awareness and motivation undermines electronic recycling efforts because it reduces the supply, thus keeping investors from putting money into recycling initiatives. In essence, people are choosing the worst environmental purchasing decision: buying new gadgets and holding on to the old ones.[31]

Because minor metals are in many diverse products, collecting and sorting them is a logistical feat that increases the costs of recycling. Most of what we customers sort to recycle never makes it to a processor that can extract and reprocess the metal, according to Barbara Reck, a research scientist who specializes in metal life cycles at Yale University's School of Forestry and Environmental Studies. Our waste management systems are ineffective; the simple truth is that most high-tech gadgets do not get recycled.[32]

The tragedy is that recycled ore often has a higher rare metal concentration than the ores found in mines. For example, only 28 tons of batteries are needed to produce 1 ton of lithium, compared to 1,250 tons of earth from Chile's lithium mines.[33] The challenge for our collection companies is to get a large quantity of material to make recycling efforts possible. In Japan an estimated 95 percent of the batteries used in the country end up in a landfill.[34]

Recycling means less mining—a huge environmental improvement—but recycling has its own waste stream because recycling is metal processing. That means waste from acids, power generation, and the transport system that hauls the recycled products from the end user to the various processing plants.

At Hitachi, workers in dark-blue pants and shirts, sporting white hard hats, masks, and yellow gloves, work on one of four disassembly lines smashing, cutting, and breaking refrigerators, air conditioners, and computers, using drills and saws. Although recycling is a complex engineering process, this first step of separating plastic from metal is primitive. "That is the easy part," Barbara Reck at Yale tells me. She then notes that recyclers focus on metals such as copper and iron, the ones used in large quantities.[35]

Few straightforward methods exist for taking apart our products to access their metals. Our products are painted, alloyed, riveted, covered in glue, and screwed shut. LCD screens have up to 250 screws in 15 different configurations.[36] Ripping them open by hand requires much labor and time, which makes it costly, especially in Japan where employee wages are high. Recently, Hitachi invented a massive tumbler to toss and break apart hard disk drives and speed up the process.[37] Workers sort through the chunks of metals and plastics and separate them into different plastic bins, which is challenging because many metals appear similar to each other. For example, a samarium-cobalt magnet dipped in nickel looks the same as a neodymium-iron-boron magnet. After removing the more commonly used metals, the workers then extract those with the highest values: gold, silver, and other precious metals. In Hitachi's case, they send that material off to a different facility. The recycling usually stops there, Reck says: "We get six or seven metals at best, on average. . . . You are still missing twenty-five, even in the best-case scenario."

Part of the reason is that no one process will extract and separate all the recyclable metals from a product; rather, its components must go through different recycling lines.[38] Often, the science is such that extracting one metal means leaving others behind. For example, once minor metals go into an aluminum or steel smelter, it is hard to recover them at all.[39] Furthermore, only a handful of smelters globally have the capacity to extract specialty metals like selenium and tellurium. And even those firms cannot get all the metals because it is infeasible to recycle all the metals from an iPhone—the best recycling facilities in the world handle only about twenty metals.[40] In fact, nearly one-third of minor metals have recycling rates lower than 1 percent.[41]

Because minor metals can be too expensive to remove from the product, they are often shredded and eventually lost.[42] The result is that only the most valuable, easiest to access elements enter the recycling stream. Metallurgists have spent centuries perfecting ways to remove metals from natural minerals. But recycling—processing complex "manmade ores"—is far more difficult. The combination of metals in products like batteries and even steels are in far more complex alloys than the finite set found in nature, which makes them impossible to recycle effectively in our current systems.[43]

Moreover, each model of a smartphone or computer, and its components, has a unique metallic composition. So recyclers of postconsumer waste do not know how much metal they can recover from a given batch of waste. Few firms want to collect material without certainty regarding its exact value. In many places, recyclers focus on only a few components or pieces for recycling.

For most products, it is just not profitable to extract minor metals.[44] Because of this, no plants in the United States, for example, recycle rare earth magnets.[45] Recycling inefficiencies are so great that we even throw away gold—recovery rates for gold are only 15–20 percent.[46] Even the world's most advanced recycler, Umicore, states that it can recover at most twenty of the about sixty metals in a given phone.[47]

When I ask Reck how green recycling really is, she pauses, "There are a lot of uncertainties around it. It depends on the metal and how easy it is to access." On one hand, she and her colleague, Thomas Graedel note, environmental savings could be far greater: "Depending on the metal and the form of scrap, recycling can save as much as a factor of ten or twenty in energy consumption."[48]

But to complicate recycling's environmental calculus, much of our high-tech wizardry contains some pretty toxic material. Some devices such as batteries are volatile and their processing risks explosion. This makes some metals cheaper to dispose of than recycle. Unfortunately, most recycling for electronic waste does not even end up in those factories but often in remote open junkyards in places like India and Ghana where people cook wires over fire, torch circuit boards, and leach out metals with potent chemicals like cyanide. The process is not only dangerous and environmentally noxious but also inefficient.[49]

The exact environmental influence also depends on the precautions taken. In many Asian countries, for example, it can be very environmentally taxing because many refiners fail to take necessary precautions such as managing exhaust. Setting up an environmentally benign and complex refining system to meet our increasingly high-tech lifestyle is many years away. As Reck notes, "For many minor metals there is not even recycling technology out yet."[50]

While recycling reduces our demand for newly mined materials, it will not eliminate it for two main reasons. First, even if we are lucky enough to collect 80 percent of e-waste, we will lose a certain percentage while dismantling and smelting, just as we lose metals in the various mining and processing stages. Even if we lose only 20 percent at each stage, we are only netting 50 percent of the metals.[51]

Second, even if we had efficient recycling systems in place, we do not have enough rare metals ready to be recycled. Many of the products that contain significant amounts of rare metals—such as wind turbines and electric vehicles—remain in use. Even by 2030, without more breakthroughs, recycling will yield less than 10 percent of demand for certain materials such as the rare earth elements.[52] Therefore, while recycling can reduce the need for new rare metals at the margin, it will not stop the mining in Jiangxi and Dalahai.

The main obstacle to the greening of our lives is that we tend to view the parts of the rare metal supply chain—mining, processing, manufacturing, use, and recycling—separately instead of seeing them as interconnected activities. The production decisions we make in one part of a product's life cycle affect the subsequent steps. By recognizing the relationship of these activities to one another, countries, companies, and even the public can do more to use resources more efficiently.

The paradox is that rare metals, although critical for our green products, have significant environmental effects. But

by having more efficient production and more effective recycling systems, the greening of our world is within our reach,
as we will see in the final chapter.

However, to reach that goal, countries need to set aside
political differences, offer incentives to make markets work
efficiently, and recognize rare metal resources as a global
good. That might be a tall order for countries because they have
become lost in the miasma of short-term geopolitical concerns.
This was clearly demonstrated at a conference in Nankang,
China, located in the midst of mining territory in Jiangxi, as
we will see in chapter 10.

X

The War over the Periodic Table

In August 2013, more than five hundred Chinese scientists and policymakers as well as a smattering of foreign researchers gathered at the Second China Rare Earth Summit in China's southern Jiangxi Province, the heart of heavy rare earth production. Officially, the conference languages were English and Chinese, but if you spoke only English you were out of luck. Just a few presentations had translations, but the interpreters stumbled over words and left long gaps of silence despite the speakers continued speech. What's more, meetings between industry and government officials were conducted only in Chinese.

While much of the rare earth research outside of China and the conferences focus on finding ways to decrease reliance on Chinese rare earths, the conference participants shared no such interest. In fact, they wanted to find ways to increase usage. One researcher even highlighted the potential for rare earths to replace rubber, a concern for China because it's the world's largest rubber importer.[1] This disparity in focus highlights the benefits of monopolizing resource production.

In the United States, there has been no scientific rare earth conference of this size in recent memory because few people study rare earths—as we learned in chapter 2, only one university even has a rare earth specialty. What's more, in 2010, the U.S. rare earth industry employed 1,500 people, down from 25,000 in the 1970s, meaning that everyone in the entire industry would have to gather to have a conference of comparable size.[2] This highlights one of the salient aspects of the war over the periodic table: its current location.

Because China and its neighbors Japan and Korea are the largest consumers of rare metals, the war over the periodic table is more pressing in Asia than in the West, where there is far less high-tech and green manufacturing even though the United States and Europe are the largest consumers of rare earth–containing products. This lack of manufacturing base is the reason why many in the West know little about these resources whereas in Japan, rare metals are front-page news. It's the same reason why corn prices loom large in Iowa, but are rarely a topic of conversation in New York.

Peering from abroad, most governments around the world see Beijing's interventions in the rare metal market as a means to control the economic spoils of future rare metal proceeds. And if you look around the host city of Nankang you can see that they have a point—signs of government largesse abound. Highway exits end abruptly at grassy knolls; six-lane roads have few vehicles on them; and unfinished high-rises, draped in netting and plastered with vacancy signs, await new occupants. But Beijing cares little about the direct proceeds from rare metals. It has greater concerns.

The regime's political legitimacy is connected to its ability to ensure that the country has the resources it needs to maintain growing high-tech and green industries that create jobs. "The rare-earth industry is quite small-scale, but the contribution to the national economy is huge," says Zhang Hongjiang, director of the Information Center at the Baotou Research Institute for Rare Earths.[3] With so much at stake for China, the geopolitical resource battles will continue.

But the reason why China dominates in research and production of rare earth elements, and indeed 40 percent of all rare metals, is not just governmental policy. It's also the result of geological good fortune and a long-term focus on developing its manufacturing base in the country. What should be worrying to resource dependent countries is that, as the value of these resources continues to rise along with concerns over increasingly dear supplies, other governments, seeking their own economic spoils and geopolitical advantage, may well follow China's lead and take a greater presence in the market.

When the Inner Mongolia University of Science and Technology flew me to Baotou for a rare earth conference in 2014, I thought the university would extend an invitation to visit Bayan Obo mine, or its tailings pond just outside the city. After all, inviting some of the leading rare earth researchers to the home of the world's largest rare earth mine without showing it to them would be like assembling chocolatiers in Hershey, Pennsylvania, and not visiting the chocolate factory. Instead of a mine visit, conference organizers arranged a trip to ride dune buggies in the desert and to see a Genghis Khan Mausoleum. The organizers said the government wouldn't allow foreign visitors to the mine.

Not only couldn't I see the mine, as a foreigner there were restrictions on what I was permitted to know. Geological data are state secrets and reliable production data are scarce. Even researching industry data can be a crime; in 2009, four employees of Rio Tinto working in China, including an Australian, were charged with stealing state business secrets for seeking iron ore data. Two years earlier, Chinese authorities convicted an American geologist of acquiring state secrets when he obtained information regarding the location of oil wells. In many other countries such data is widely available and, to many in the industry, this data seemed commercial. Authorities held him until 2015.[4]

China's secrecy and a heavy hand in the market undermine the ability for the world to meet its resource needs efficiently. Few want to invest in an industry where market data are unknown and the government can quickly change the rules of the game. Although this situation thwarts investment, it plays into Beijing's long-term strategy.

Beijing uses access to their rare metals as a way to attract foreign investment that will subsequently bring international technology to China. The central government wants to upgrade its businesses. In a sense, Beijing is modeling its industrial growth after Japan's Hitachi Corporation—once a mining services company, now it's one of the world's largest technology companies.

From the outside, China's central government may appear to be in sole control of the rare metal industry. It has a guiding vision and even white papers written that outline its future goals. But the reality is far different. Central authorities face a formidable challenge in implementing policies because numerous government bodies are vying for power; the industry is rife with corruption and a divide exists between central and regional officials. Part of the problem stems from a government decision in the late 1990s to abandon its Ministry of Mining and Geology so as to streamline the bureaucracy. But instead of simplifying policymaking, this jettisoning of the mining ministry devolved responsibility and left authorities with a hodgepodge of agencies, of which about thirteen have significant influence over the rare metal market but without complete control. Each vies for power.

What's more, Beijing holds surprisingly little sway in remote mining areas. To understand Beijing's complex relationship with those regions, one would do well to heed the Chinese adage, "A strong dragon cannot beat the local snake." Surprisingly, on many policy issues the local authorities are far more powerful, a fact the central government concedes. In 2012, the State Council admitted that it lacked oversight of rare earths production in the regions, and reported, "There are so many mines scattering over a large area that it is difficult and costly to monitor their operation." The head of the Rare Earth Office of the powerful Ministry of Industry and Information Technology also commented that the high profits from illegal mining have led to an alliance between local government officials and illegal miners.[5] In many cases, local officials see little reason to follow Beijing's diktats, such as to shut down mines, raise environmental standards, or reduce output. In 2002, they thwarted Beijing's attempts to consolidate the rare earth industry into two companies to meet national goals. Regional interests are more parochial—economic development and prosperous business. And since local officials also control the appointment of judges, they have outsized influence.[6] In a show of defiance against Beijing's plans, the vice secretary of the Jiangxi Rare Earth Association said, when discussing whether a state-owned company would buy local firms, "in the face of state policies, we will walk our own path."

But it's not just local governments that thwart Beijing's strategy goals. Companies themselves, both state-owned and private, often prefer to sell the metals for immediate gain. The incentive to sell the resources, even if it means violating policy, contributes to the black market.

With so little control, Beijing cannot rely solely on regulations because they often backfire. For example, a researcher at a Chinese rare earth association relayed to me an example of the unintended consequences of new rules. In 2012, Beijing issued a regulation aimed at reducing production overcapacity, stating that all rare earth refiners with a production capacity below 5,000 tons must close. Instead of shutting down, many smaller refiners increased their capacity to more than 5,000 tons, thereby increasing the country's production capacity. By 2013, the rare earth sector had the ability to produce 300,000 tons per year, even though the world consumes less than half that.[7]

Beijing's policy is rarely enforced efficiently; paradoxically, it doesn't have to be enforced evenly to meet its desired ends of developing high-tech companies on its own shore. Beijing just needs to introduce enough uncertainty into a foreign high-tech company's supply chain to convince jittery executives that it's easier to manufacture in China than outside of it. It's akin to Beijing's strategy for encouraging its citizens to use domestic search engines instead of Google: if the authorities interfere with Google enough that the site works haltingly and with little reliability, people won't use it.

But as the largest consumer of rare metals (and indeed numerous resources), China has its own resource concerns. Authorities often feel at a disadvantage with international mining and trading firms that have pricing power. They feel they have been taken advantage of; country officials point to the oligopolistic control of the iron ore market, where they feel they are price takers from the three companies that control 70 percent of the trade.[8]

Despite China's own resource concerns and its inability to completely control its own market, most other countries frame their minor metal resource strategies around China's dominant position. And they are now worried.

When China's economy was far smaller and before its accession to the World Trade Organization (WTO) in 2001, few countries fretted about the socialist government's control over its industry. That changed when China joined the WTO and had to adhere to international trade practices. Despite China's enthusiasm to join the international trading community, its relationship with the organization is an uneasy one, built on awkward footing because China's free market is guided by the Party first, and the law second.

The leading advocate for WTO membership at the time, then prime minister, Zhu Rongji, warned Party members four months after accession, "Western hostile forces are continuing to promote their strategy of Westernizing and breaking up our country."[9] So although the country welcomed the economic benefits of additional trade, it looked wearily on the framework that brought it.

Leadership feared that the international trading system could upend the Communist Party's role. The challenge to the current order, as Teng Biao, a Chinese scholar formerly at Harvard and a critic of the government, told the *New York Times* in 2014, is that "The basic political system is incompatible with rule of law," a key component of a rules-based trading system. He argued that friction comes between Chinese Party leaders who mainly want to use the law to control society and its people rather than see it as a guiding, governing framework.[10]

For more than a decade China has tried to balance a socialist system where the government is the arbiter of the laws with the demands of a rule-based international trading system. Over that time, global trade flourished, foreign companies invested hundreds of billions of dollars and the country continued to develop a rare metal resource base to support growing industry. But times are changing.

Beijing has become confident with its increased economic prowess and savvier with its interpretation of international trade rules. Terms like "equal access," "antimonopoly," and "sustainable resource development" are foreign concepts in China. Its leaders interpret these terms in their own context, through a lens of state control of the market. The terms may mimic Western thought, but they are far removed from their cultural significance and original meaning, like Westerners who adorn their walls with or get tattoos of ornate Chinese characters.

For example, in 2014, many foreign companies and chambers of commerce complained that Beijing was using antimonopoly regulations not to break up true monopolies but as a weapon. They felt the government was using the pretext of monopolistic regulations to reduce the role of foreign firms in sensitive sectors.[11]

Similarly when the United States accused China of using export quotas on rare earths and a slew of other minor metals to forward its industrial policy in the WTO, Chinese officials claimed they had used them to conserve exhaustible natural resources and protect the environment. But there was little mention of the environment in the press when China first instituted quotas in 1999 for rare earths. Instead, China's export and production controls were devised to rein in price competition, which sent rare earth prices downward. (During that same year, the main rare earth companies came together to set a minimum price.[12])

But the export controls became more than that. As the vice chairman of the Inner Mongolia Autonomous Region said of the regulations, "We are not taking the short-term view of just trying to prop up prices. Imposing controls and reducing exports aim to attract more factories using REEs [rare earth elements] from home and abroad to Inner Mongolia."[13]

In certain ways, many in Beijing resent producing nearly all the world's rare earths from some of the most polluting mining and processing systems in the world, and selling them overseas. From the Chinese perspective, countries have outsourced their pollution to it. And they are correct because a good portion of the rare metal that China refines could not be refined elsewhere like in Japan due to environmental regulations. Some Beijing officials believe they have sacrificed the country's environment for little profit.

Instead, Beijing wanted to see much more economic gain and by 2010 took more extreme measures. It dropped the country's rare earth export quota by around 40 percent during the year.[14] The move sparked backlash from consumers that ultimately led to two WTO resource cases, which China lost.

The WTO decision outlawing China's export controls changed China's tactics, but not the country's strategy of using domestic resources to promote growth. Over the past five years, Beijing has begun consolidating control of the rare earth industry—once dominated by hundreds of mostly small operators—into large state-controlled firms. Consolidation may not prove as effective as export controls, but with the production of rare earths now in the hands of state players, China has effectively strengthened government control over the industry. Consolidation may also help to stem smuggling, which, by some estimates, accounts for 40 percent of the country's exports.[15]

Beijing has additional policy tools to control the minor metal market. It provides subsidies and below-market-rate loans to state-backed mining firms; purchases unsold rare metals to stockpile; and uses complicated export licensing processes to restrict competition.[16] These policy tools give countries concern about the future of rare metal supplies. As the U.S. Congressional research arm notes in a 2015 critical materials report, "China is likely entering an era of fewer raw material exports over the long run, which requires some type of long-term planning."[17]

Although many countries feel that China is overextending its control of the rare metal market, the sense in China is quite different. Qinhua Wang, the vice chairman of the China Nonferrous Metals Industry Association, noted, "We have not enough influence on rare metal in the world and we didn't get reasonable profit." He later added he expected China to reorganize, to better capture rising prices, and to have larger international market influence.[18] Much of that strategy relies on importing foreign technology, which makes Japanese trade officials nervous.

To defend against a migration of Japanese tech corporations to China, Tokyo officials handed out grants of up to $70 million each to companies for programs related to conserving rare metals and rare earths, ostensibly as an inducement to keep manufacturing facilities in Japan. But as officials know, financial incentives and arm-twisting can only go so far. Business interests eventually prevail. As one Japanese trade official commented to me, "It is impossible to keep one technology in Japan forever."[19]

For Japan, much is at stake. The country spent well over $1 billion to reduce its reliance on Chinese rare metals in 2011 alone. As an island nation reliant on imported materials, Japan learned decades ago that resilient resource supply lines are critical to a stable economy and to regional security.

A little history is in order. In the first part of the twentieth century, the archipelago's quest for natural resources to feed a growing, resource-scarce country led to conflict throughout the Asia-Pacific. The country's army reached into China, Korea, Indonesia, and Burma (now Myanmar) for supplies, sparking conflict as the world entered World War II.

The high stakes rare earth battle in 2010 discussed in chapter 2 has amplified Japanese officials' concern about their country's economic security in several dimensions. First, they worry China will use trade in rare metals as a geopolitical weapon and again cut exports. Second, they fear the continuing exodus of their companies to China as the country offers unfettered access to lower-priced resources. Third, once in China, Japanese officials fear they will lose their technology either through theft or lack of domestic innovation. This will enable China to compete directly with Japanese exports. That would be a big blow to Japan's economy. For example, the country's electronics exports total about $120 billion, historically accounting for about 15–20 percent of its shipments abroad. "Our industries depend on these materials. It is an issue of competitiveness," one official told me.[20]

So in that sense even with prices now far lower than they were in 2011 and with supplies seemingly abundant, the insecurity over future resources still concerns many Japanese officials and companies. And without domestic resources, they feel they are in a bind. As another official commented, "We have no reliable alternative for heavy rare earths but China." The same is true for tungsten, indium, and antimony.

To avoid overreliance on China, Japan's rare metal strategy goals are to increase the availability of sustainable rare metal supply lines globally and wean companies off those rare metals that have the greatest supply risks. The plan is based on four pillars: securing resources overseas through investment and diplomacy; recycling; developing alternative materials; and stockpiling.

Tokyo has provided firms like Masato Sagawa's Intermetallics millions of dollars to develop ways to remove dysprosium from magnets and has provided money to universities to educate a cadre of mineral scientists and engineers. Officials also loaned $325 million to Lynas Corporation in 2011 to develop its rare earth producer with facilities in Malaysia and Australia.[21]

Japan's efforts to fund the development of new mines and to become more efficient at recycling will add more supply, but it's unclear whether their efforts alone will lead to increased resource security. They are slow steps and even if, for example, Japan succeeds in its recycling goals, recycling rare earths will meet only 10 percent of the country's demand by 2025.[22]

What's more, it's unclear whether some of the country's goals would be effective in a time of crisis. For example, Japan has made it a national goal to ensure that Japanese trading companies physically import at least 50 percent of rare metals. But this government policy is "meaningless," one Japanese manufacturer commented, because Japanese companies will act in the best business interest of the company, not necessarily the country.[23]

While Japan has, by far, the most well-funded rare metal security strategy, other countries have made similar efforts, albeit with different foci. The European Union focuses on writing studies and reducing trade barriers in the rare metal sector. Meanwhile, the United States, also concerned about fair trade, is far more focused on providing an unencumbered supply of rare metals for its military. Many countries stockpile metals for economic reasons, but the U.S. military stockpiles them for defense purposes.

The true challenge lies in assessing the effectiveness of these strategies because most of them are long term—the effects of research and trade policy adjustment take many years to evaluate. But it's hard to imagine that tens of million dollars a year in research and a few stockpiles can insulate countries against the myriad of risks. It's also hard to believe that these countries look at ensuring resource supplies as a risk. After all, as we noted in chapter 8, the U.S. government allocated nearly $250 million for titanium research in the 1950s. That's more than the country has allocated to all rare metals over the past decade.

Over the past ten years, most major economies and many resource-focused think tanks have published criticality studies examining resource constraints for rare metals. While individual assumptions and methodologies vary, most look at critical risk in two dimensions, the importance of a particular material to an economy and the likelihood of its supply disruption.

Combining all the risk lists from each report published over the past decade reveals that more than half the elements on the periodic table are "critical" to one country or another.[24] The United States fears shortages of terbium, the European

Union thinks antimony is at risk, and Britain frets over tungsten, the metal they once sold to Germany for armaments during World War I. Depending on the author of a given study, only a handful of materials or 40 percent of them might be deemed at risk. These differing results are based on a study's design and the author's assumptions of how to assess the risk. And they vary drastically, offering little guidance to policy makers and company executives.

Some researchers fear supply shortages of metals like tellurium, which face a scarcity risk; others see niobium facing economic or market risk, and dysprosium faces a "lack-of-substitute" risk.[25] But how realistic can these assessments be?

Despite the variety of report perspectives, the studies fail to help generalize which metals are facing shortfalls. As one critique of these studies commented, "the methodology is immature and the results are not necessarily helpful to all parties whose ultimate aim is to secure future supplies of minerals."[26] Much of this is because of a lack of data in terms of both material supply and demand, as discussed in previous chapters.

The U.S. Department of Energy notes, in its *Critical Materials Strategy* report, that it could not use traditional economic models or analysis to predict the supply and demand of materials crucial to the clean energy world because of the difficulties of predicting unforeseen technological breakthroughs.[27] It is difficult, if not impossible to calculate the type and number of wind turbines, solar panels, and tidal power systems that will be produced and the amount of resources each will consume two decades into the future. In fact, research institutions have been historically poor at predicting far more straightforward resource demands such as the demand for coal.

Coal demand should be predictable because coal is largely used to produce power. Demand is a function of the capacity of all coal-burning facilities and predicted economic expansion—the faster the growth, the more coal consumed. Despite the simplicity of predicting coal demand, the International Energy Agency (IEA) drastically underestimated coal use in the first decade of the 2000s. In 2004, the IEA projected that the world coal consumption rate would grow 1.4 percent annually to reach about 3,600 metric tons of oil equivalent (a measurement that allows comparisons between different energy sources) by 2030. However by 2011, the world was burning 3,700 metric tons of oil equivalent of the black stuff because of the growth of China and India. If the legions of experts at the IEA cannot foresee coal demand, predicting each minor metal's demand will be guesswork, at best, because far less is known about these materials, and even less about where and how they are mined and processed.[28]

Robert Jaffe, a Massachusetts Institute of Technology (MIT) theoretical physicist, who was the co-chair of a 2011 joint American Physical Society and Materials Research Society report, *Securing Materials for Emerging Technologies,* has difficulty convincing skeptical lawmakers that the United States does not need to be mineral independent. Is the United States coffee-, banana-, or even saffron-independent? Jaffe asked me. "For us to be independent in all these panoply of foods is crazy," he told me. Adding that it's expensive because "bananas don't really grow here."[29]

He has a point. The United States imports 86 percent of all the seafood it consumes, despite controlling more of the world's oceans than any other country.[30] But people in the United States aren't clamoring for fish independence. As long as the country has reliable and inexpensive seafood supplies, does it matter where they come from? So why do we have to be metal independent, Jaffe wonders. It's a question without a clear answer.

It's not that Jaffe is unconcerned about future supplies of rare metals. When the rare earth crisis occurred in 2011, he referred to rare earths as the "flavor of the month" because of the furor focused only on this one set of rare metals. Jaffe was just as troubled about the supplies of a slew of other critical materials including tellurium, germanium, and lithium. "A host of other elements are poised to present problems in the future," Jaffe testified before Congress in 2011.[31]

Jaffe feels the notion that the United States can mine itself into resource security is absurd. But he is fighting against the long-held American populist obsession with resource independence.[32] The sense of U.S. self-reliance harks back to explorers heading West in search of natural resources, from gold to coal, when the country had an abundance of the resources it needed. It wasn't until the early part of the past century that Americans began to acquire resources from abroad.[33]

The United States still has a self-sufficiency tendency as well as relative abundance compared to other countries, but it has a stubborn faith in the free market. The market allocates resources and capital; the government is not supposed to pick winners. This is why, in 2011, when Solyndra, a solar panel manufacturer, went bankrupt, even after receiving federal government subsidies, President Obama faced a backlash from the conservative wing of Congress.[34] A stark choice challenges U.S. politicians—have resource self-sufficiency or a free market.

Without government intervention, investment dollars will flow to the mineral deposit with the greatest potential for production. For many metals, those deposits are not in the

United States. To become self-sufficient is a costly endeavor, as Jaffe notes. But for those in Congress who are more concerned about national security than economics, his arguments fall on deaf ears.

Although the U.S. military uses limited quantities of rare metals—only about 5 percent of all rare earths consumed in the United States—military needs are the main driver of the U.S. resource policy debate. "There is no consensus of what is a strategic and critical material. But we are all well aware of the behavior of China," one staff member of the Committee on Natural Resources told me.[35] He added, "It's hard to get mineral issues on the radar of Congress if there is no security or critical angle."

Instead of passing legislation on natural resource security, Congress uses the concern over rare metal supply as a way to rehash long-running mining, environmental, and land use battles. In fact, Congress has been loath even to define what critical materials are. Committee chairman "Doc" Hastings noted that such a definition would create a scenario where basic materials like sand and gravel, hardly strategic, would be unavailable for emergency construction.

With little congressional action, the government has made modest efforts to increase research on the development of alternative materials to replace rare metals, to enhance market research on rare metal markets, and to continue to add to the defense stockpile.[36] Despite the jockeying and the cursory efforts to ensure rare metal resource access, many of the efforts miss the larger picture. With so many rare metals, refined in so many grades that go into numerous components, ensuring rare metal supply is complex.

The world needs more resilient minor metal supply lines to meet growing needs. Steps to reduce reliance on rare metals cede opportunities to countries where the rare metal resource supplies are secure. And that has profound economic implications.

Jaffe's report warned that a shortage of critical metals "could significantly inhibit the adoption of otherwise game-changing energy technologies. This, in turn, would limit the competitiveness of U.S. industries and the domestic scientific enterprise and, eventually, diminish the quality of life in the United States." But the dangers are more profound on the long-term economic health of a nation.

Three researchers from Carnegie Mellon University have shown that U.S. innovation of rare earth magnets dropped when U.S. rare earth industries shifted to China.[37] Cut off from the supply of raw materials, U.S. researchers were less likely to continue development on rare earth material science. Meanwhile, Asian countries increased their expertise.

The shifting of research jobs to countries where the most technologically promising materials exist should not be underestimated. It is a large economic concern because rare metals are at the heart of future sustainable industries like green technology.

About 4 percent of all U.S. patents issued between 1986 and 2013 involve rare earths. Adding the number of patents that are associated with molybdenum and tungsten, two other minor metals involved in a 2012 U.S.-led trade case that China lost, the number roughly doubles.[38] With so few employees in the section, the United States is losing the expertise to develop its own resources and allowing future technological advances to go to other countries.

U.S. universities are also not minting the number of scientists that the Rare Metal Age demands. Luka Erceg, founder of Simbol Materials, a lithium extraction company, stated that because no university in the United States offers geothermal energy degrees, it has taken him nearly a year to find qualified candidates.[39]

The United States has already witnessed the same fall in stature in the nuclear industry. Once a global leader in technology, it is now struggling to regain the position it lost in the early 1980s.[40] Japanese officials have the same concern. One lamented that the country was losing its expertise in glass polishing, which is crucial to high-end lenses, as firms leave Japan out of concern for rare earth metal shortages. Japan responded quickly to China's rare earth export ban in 2010, because Japan is the world's largest rare earth import market because of its high-tech and heavy industries. Conversely, the current lack of high-tech manufacturing in the United States appears to act as insulation from China's resource strength.

This could change quickly. The United States is beginning to bring manufacturing back within its borders, or to "reshore." Take Apple's 2013 decision to begin producing computers at a Texas plant for the first time since 2004. Other companies such as General Electric and Caterpillar have followed suit. In fact, a Boston Consulting Group study in 2014 revealed that 16 percent of large companies are bringing manufacturing back to the United States, more than a doubling in two years. As more tech manufacturing returns to the United States and green tech companies flourish within its borders, Washington, like Tokyo, will over time wake up to the increasing importance as well as the perilous state of rare metal supply lines.[41]

The policy response to the increasing demands for rare metal is not straightforward, nor is it simply to promote resource production at home, at any cost. As Jaffe's report notes, import reliance is beneficial to the United States "if foreign sources are diverse in number and location and can supply the elements at a lower cost than domestic alternatives." The United States must remain diligent, the report concludes, because if just a few of these green energy applications take off, the geologic scarcity, the long lead times to develop supply lines, and the complicated methods of processing rare metals could mean there won't be enough of them to meet the demand.

"History is on the side of finding new supplies," Jaffe tells me, referring to the fact that investment money goes into finding new deposits when metal prices are high, "but I'm very concerned that this argument may fail for these materials."

That is, unless we have a plan.

XI

How to Prosper in the Rare Metal Age

In 1970, yellow legal pads, typewriters, and double-stacked in- and out-boxes covered desktops, not icons. Although computers had yet to make their appearance on the average desk, a resource change was coming. Centronics introduced the dot matrix printer, Intel began selling microchips, and programmers had just sent the first e-mail message on ARPANET (Advanced Research Projects Agency Network), the roots of today's Internet. A few years later, George Pake, who headed Xerox Corporation's Research Center, predicted that by 1995, TVs attached to keyboards on office desks would strike the death knell for paper. "I'll be able to call up documents from my files on the screen, or by pressing a button," he said. "I can get my mail or any messages. I don't know how much hard copy [printed paper] I'll want." As it turns out, he'd want much more.[1]

Despite all the high-tech gadgets that appear to negate the need for paper, paper use in America has nearly doubled since Pake's days. We now consume more paper than ever:

400 million tons globally and growing. That's roughly 2 pounds of paper per office worker every day.[2]

Paper is not the only resource we are using more of. Technological advances often come with the promise of using fewer materials, but the reality is that they have historically driven more materials use, making us reliant on more natural resources. The world now consumes far more "stuff" than it ever has. We use thirty-four times more construction minerals such as stone and cement and twenty-seven times more ore and industrial minerals, such as gold, copper, and rare metals, than we did just over a century ago. We also each individually use more resources. A person today consumes more than ten times the amount of minerals than one did at the turn of the twentieth century.[3] Much of that is due to our high-tech lifestyle.

In the 1980s, Americans huddled around a TV, maybe an Atari game system. Today entertainment flat-panel systems are hooked up to DVRs, speakers, and game systems replete with cameras and motion sensors. The cassette and record players of past years have been relegated to the antique stores, but many new products complement existing ones rather than rendering them obsolete. The microwave revolutionized the Western kitchen but it did not replace the stove, oven, or grill. Likewise, since the tablet computer and smartphone sync with each other, we find them complementary despite their similar functionality. Americans seem to have little trouble with this redundancy: even in 2013, more than a third of Americans owned a tablet, laptop, and smartphone.[4] Meaning Americans are purchasing a lot of gadgets. According to Consumer Electronics Association, the average U.S. family owns twenty-eight different electronic devices not including kitchen appliances, power tools, or washing machines.[5]

For proof of this consumption pattern globally, we need to look no further than our trash. The amount of electronic waste the world is producing is growing at an estimated 17 percent annually, even though total amount of waste collection in some countries has leveled.[6] We are on a global trajectory to toss out over a billion computers annually. This is not just because we have more of these devices but because we use them so briefly. The average lifecycle of a smartphone is about twenty-one months. Likewise, laptops, tablets, and many of our high-tech gadgets have life spans of less than three years. This is not because the product is useless when we junk it but because its obsolescence, in many cases, is by design.[7]

iFixit, a site dedicated to repairing those irreparable high-tech gadgets, notes, "Apple is making billions by selling us hardware with a built-in death clock," referring to a built-in battery that iFixit believes starts to lose power just as the warranty ends. To replace the smartphone's battery, you must either go to an Apple store or mail the phone to Apple along with $80. As few people are willing to forgo their phone for the time it takes to repair, they are more likely to upgrade. American mobile telephone companies also institutionalize minor metal consumption by offering to provide a new, "free" phone as frequently as every six months.[8] Because of this, product life cycles are now measured in months not years. And this has a profound effect on our resource use.

According to a Japanese industry study, in the 1970s the average commodity had a life cycle of five years and about 80 percent of all the materials in them had a life cycle of at least three years. Just after the turn of the millennium, that dynamic nearly flipped. Twenty percent of all commodities were discarded in one year, and half, in less than two years.[9]

Today consumers also find it far easier to buy many high-tech goods because by comparison they are cheaper, in some cases drastically cheaper compared to other items such as food. Since the early 1980s, the consumer price of a television and other video equipment fell by more than 90 percent. But these trends don't even address what might be the biggest use of rare metal resources: the technology of the future.[10]

With a sprinkling of rare metals, Robert Tenent, senior scientist at the National Renewable Energy Laboratory in Golden, Colorado, turns no-tech glass into high-tech windows. Tenent's windows allow the sun's light to shine through, but not its heat, or alternatively, the sun's heat, but not its light. Sometimes the windows keep both out—flip a switch and they darken within five minutes. The windows are part-curtain, part-insulation, and undoubtedly contain advanced technology. Tenent's secret ingredients are a few grams of tungsten and indium.[11]

His windows are part of the laboratory's showcase building—an of office of 33,445 square meters that uses no external electricity, so keeping cold in during the summer and heat in during the winter is crucial. Tenent's windows are not just good for the lab; they present a great green opportunity: replacing old energy-inefficient windows can save up to 4 percent of all energy consumed in the United States. But at the same time, the windows present a tremendous challenge for our rare metal supply lines because the United States alone has more than 1,813 square kilometers of traditional windows. The windows mark the next stage in the Rare Metal Age when demand soars: when the infrastructure that underpins our modern lives becomes high-tech—fleets of electric buses, roads built out of solar panels, or elevators that rely on magnets instead of cables. Michael Silver, the chief executive officer of American Elements, a material science company, tells me, "The spigot [of ideas] is on full blast. You start seeing astronomical volumes of material ten years out."[12]

New innovations include not only new products but also existing technologies used in new ways. Imagine the interior walls of a bedroom lined with flat panels that change color to adjust your mood. Or bathroom "mirrors" that display your body's vital signs gathered from sensors and cameras around the house. General Electric foresees a kitchen designed around a hub with a cooking surface that uses voice, motion, and facial recognition to help you share your culinary creations with others via the Internet.[13]

These products may seem far off, but for some perspective on how quickly new technologies can proliferate, look back at the smartphone. In just four years after its introduction, 6 percent of the world population owned one, making such phones, by some metrics, the fastest growing technology ever. A few years later, the tablet accomplished the same feat in half the time and now nearly half of all Americans own one. This rate of technological penetration is the new norm and it drives up rare metal use. Cisco, the American network equipment company, reports that in 2010 over 12.5 billion devices were connected to the Internet. That number will quadruple by 2020 to 50 billion. Interconnectivity will drive rare metal demand, not just for the products themselves, but also for the infrastructure to power them. This is especially true if it's new green technology.[14]

At the same time that the wealthy world is using more rare metals, the speed at which developing countries are playing technological catch up is unprecedented. For example, in 1995, only 7 percent of Chinese city dwellers owned refrigerators, twelve years later, 95 percent of them did.[15]

The U.S. government's National Intelligence Council predicts that the global middle class may nearly triple over the next two decades, adding roughly two billion more people— the equivalent population of roughly two more Chinas or six Americas. "Such an explosion will mean a scramble for raw materials and manufactured goods," a 2012 council report says. Add to this our global quest to find green energy alternatives, and the demand for rare metals will soar. The Japan Institute of Metals, meanwhile, reports that demand for rare metals like cobalt, tungsten, and lithium will increase by a factor of five by 2050, and will outpace our current reserves for many of them.[16]

The coming resource crunch raises the likelihood that resource-rich countries will use their own increasingly valuable resources to gain strategic and economic advantage. This portends tense showdowns between individual companies and countries as these countries continue to tighten their control over the metals, or worse, cut rare metal trade again. All this resource use scares Roger Agnelli, former president and CEO of the mining giant Vale. As he told me in 2013, "The reality is the planet is very small for the number of inhabitants we will see in 2025. As technology is getting cheaper, resource demands are increasing and we are facing changes in geopolitics. This is real."[17]

And yet, I would argue, the answer to our concerns about rare metals is not to shy away from using them because of our geopolitical supply fears. Rather it is to search for more sources, use them more efficiently, and advance our knowledge of geology, metallurgy, and material science.

To deal with potential resources shortages, we must think about minor metal supply in several dimensions. The world, and indeed each country, needs a secure supply of sufficient resources, at minimal environmental cost, arriving through resilient supply lines. Therefore, we need to turn rare metals into commodities—we should strive to make them cheaper, more abundant and produced with the least environmental impact. That means international efforts are critical, including to develop, understand, and improve material development and flows; to invest in education regarding the use and conservation of resources; and to adjust regulations to better govern the start-up of new mining operations.

In the Rare Metal Age, we either need to get better at predicting the future or set up robust supply chains for a variety of rare metals that could possibly be in demand (or both). But the track record of our experts in predicting the technological future is muddy at best.

- "Television won't be able to hold on to any market it captures after the first six months. People will soon get tired of staring at a plywood box," Darryl Zanuck, founder of Twentieth Century Pictures (1946).[18]
- "There is no reason anyone would want a computer in their home," Ken Olsen, founder of Digital Equipment Corporation (1977).[19]
- "I predict the Internet . . . will soon go spectacularly supernova and in 1996 catastrophically collapse," Robert Metcalfe, founder of 3Com and the Ethernet (1995).[20]

Since we don't know which invention will take off, we can't estimate which rare metal will either. Thirty years ago, dysprosium had little use. Now, in part because of its use in magnets, it is essential for our new high-tech lives. Gallium, because of its low melting point, could find itself in high demand in 3D printing, a type of home-based manufacturing. Or gadolinium, a sister element to dysprosium, has long shown promise in a magnet to produce energy-efficient cooling. This future technology could revolutionize the refrigerator market, putting that appliance within reach of the billions who do not now have one. Or it may always be a technology of the future.[21]

Rather than predicting the future, we should prepare for it. And for that we need to mint more people like Toru Okabe. Okabe, a Tokyo University material science professor, is so enamored with titanium that he gives it out as gifts. When we met at a conference in southern China, he handed me a small sealed bag with a thin titanium coil to explain the concept of shape memory. (He told me that if I stretched out the coil, it would return to its original shape, which makes titanium useful in glass frames, for example.) Titanium is not just a material science toy, it has an abundance of uses and if it were less expensive it would transform our resource demands because titanium is stronger than steel, 45 percent lighter, and corrosion resistant.[22]

The challenge involved in titanium, the fourth most common metal in the earth's crust, is that it's expensive to produce because of the high temperatures needed in processing, making it environmentally taxing. Okabe wants to improve titanium's processing efficiency and turn it into an abundant commodity. If he succeeds, a fraction of titanium could replace steel in bridges, buildings, and even tools. Titanium could be the green alternative. Products that use titanium instead of steel would require less metal, which means less mining and greatly reduced carbon dioxide emissions.

Okabe has a leg up on researchers elsewhere. He lives in Japan where his university receives a modicum of government support, whereas most metallurgists elsewhere have been less fortunate. In the United States, since the government shuttered the Bureau of Mines in 1995, Colorado School of Mines professor Patrick Taylor has had almost no way of securing government funding to support extractive metallurgy research. So he must increasingly work with industry from overseas.[23] Funding is crucial, but as the industry is focused on the bottom line, corporate funders are not interested in experimental research, the kind of work that pushes frontiers, as Taylor tells me—the research that can truly revolutionize the industry. For Taylor, U.S. government support would be a boon. Furthermore, studies show that government research spending in physics and chemistry is the best predictor of economic growth in many countries and can yield up to $10 for every $1 invested in materials research.[24]

Beyond economic growth, the world needs more university-level mining, metallurgy, and material science programs to help alleviate the shortage of mining and metals professionals. They are needed to create the scientific breakthroughs that the world demands. Increased government support is crucial, but it is not enough.

Unfortunately, our best and brightest who have material science and physics degrees aren't conducting research. They head off to Silicon Valley or Wall Street, Elisa Alonso, a former Massachusetts Institute of Technology material scientist, tells me. When I ask her whether many of her colleagues will stay in material science after nearly a decade of study, she laughs, "You are not going to do that." The other jobs being offered are too exciting and lucrative to turn down.[25] We need to bring prestige and romance to toiling with metals and to start companies that ask big questions, which only advances in material science can answer: for example, how to build a more fuel efficient car and commercialize space travel.

Visionaries like Elon Musk, the cofounder of PayPal who was admitted into Stanford's material scientist doctoral program, has started companies that ask just those questions, Tesla and Space-X. We need more of them. Simply, we need to create excitement around material science as we have for entrepreneurship. Now 70 percent of millennials want to work in more entrepreneurial endeavors outside a corporate structure. We need that same level of excitement for science so Alonso's colleagues will stay in the field. They are not only critical to finding new breakthroughs with rare metals, but also in developing ways to be more efficient with the metals we currently use, as well as discovering more abundant and greener materials that can limit our growing dependence on certain limited resources.[26]

While research is crucial to unlock material science secrets, we also must change consumer habits and business models to ensure sustainable supplies of resources. Therefore, we need to fundamentally change our relationship with our gadgets; we cannot continue to buy a new smartphone when the battery dies. Our gadgets must last longer. Changing a cracked smartphone screen must become as simple as changing a battery in a remote control. High-tech repair services need to be as ubiquitous as dry cleaners. Companies like Apple, which make lengthening the life of its products difficult, must play a role in prolonging the lifecycle of its products by selling component replacements and providing easy access to materials under the hood. In essence, it needs to open its products' ecosystem to help save our planet's.

Selling fewer products that last longer may sound as if it is anathema to a market economy; it's not doom and gloom. For Caterpillar, which began a remanufacturing program in

1972, taking back equipment is a core part of its business. In fact, remanufacturing helped the company to open markets in the developing world and provided new profits because re-manufactured products are less costly to produce and they sell for up to 60 percent of the price of a new tractor. Companies could also lease products much the way car dealerships do, taking the product back after use. They need to put as much effort into the end-of-life care of their products as they do into its functions.[27]

Governments need to hold companies responsible for the effects of their products. One company that recycles millions of phones is easier and creates more profit opportunity than would be achieved by asking millions of people to recycle one phone each. Not only is it more efficient, it creates recycling feedstock and economies of scale that help create a profitable recycling supply chain.

Regulators would do well to require companies to include afterlife care as part of the products they make as Dell Corporation does. Mandating that companies must recycle their own products encourages companies to more efficiently use materials—especially rare metals; to design products that are easier to recycle or reuse; and to develop less environmentally harmful and more easily recycled alloys. Efficient recycling will extend the life cycle of rare metals and reduce the need to dig for more.[28] But only one quarter of U.S. electronics end up in effective recycling systems. Now, in places like South Portland, Maine, companies are already mining landfills for valuable metals the previous generation junked in the 1970s. We must not set up the same dynamic by tossing away rare metal–laced gadgets.[29]

One step to improving minor metal resource efficiency could be as simple as labeling. For people to begin to make educated decisions about their resource consumption, they need to know what they are consuming. Labeling also encourages companies to know what materials are in their products and can facilitate recycling when the information is placed in bar codes on the side of a product.[30]

A report by the Ellen MacArthur Foundation states that a circular economy, based on remanufacturing, reuse, and recycling would save $1 trillion in material costs by 2025 and create one million jobs in Europe alone. It could also fuel the next generation of recycling techniques, which the world desperately needs.[31]

An even better way to keep minor metal resource use in check is decidedly low-tech: conservation. Energy conservation measures—such as improving the efficiency of energy grids, electronics, and buildings—reduce the need to build or expand infrastructure that relies on minor metals. For example, it's far less resource intensive to add extra insulation to reduce energy demand than to install a solar panel to produce more efficient energy; it's also a good way to consume fewer rare metals.[32]

While the answer for resource security won't come solely from government capitals, the right legislation and regulatory changes can have a great influence. In addition to extending producer responsibility, providing greater funds for technological innovation and adopting energy conservation measures, governments should develop long-term plans to increase rare metal supply. A good start for countries with resources would be to clarify the time frame and review process for new mining operations and specify the places that are off limits to exploration. With the length of time required to open a mine extending over a decade, governments should examine ways to speed up an environmentally sound development process. Wealthy governments should not outsource pollution.

Another role for government is to help produce and disseminate market data to solve market problems discussed in chapter 5. This is a role that the U.S. Geological Survey helps to fill, although cuts in funding over the years have adversely affected the consistency of the information. Such research can help to identify gaps in supply and anticipate future trends. Governments should also work more closely with research institutions and the tech industry to better understand future demands for rare metals.

Governments can help to encourage production by offering tax incentives; subsidized insurance to reduce risks for mining investments; and by taking debt or equity stakes in companies. Less traditional incentives, such as purchasing agreements, can ensure production and investment during periods of volatile pricing. And for start-ups like Avalon Rare Metals in Canada, government support and broad statements of policy support could be the boost needed to secure long-term investor funding. But most governments outside of Asia are less likely to "pick" winners.

Some studies argue that stockpiling rare metals is a key government policy tool to ensure material in times of shortages. However, stockpiling rare metals is unlikely to be successful because manufacturers need a variety of grades of material, making it nearly impossible to stock enough material in the right grades. What's more, if a stockpiling country lacks processing facilities, stockpiling a rare metal is like stockpiling cans of tuna without a can opener: well-intentioned, but useless. Stockpiling may well be very expensive because governments often buy when prices are high, forcing prices even higher.

What may be more useful, although less politically palatable, is for governments to set regulations or offer incentives for companies to stockpile the rare metals they require and the components made from them. Companies know what their needs are and therefore have an advantage over governments in choosing which materials to stockpile. But as we saw in chapter 6, companies are reticent to tie their cash up in resources that may fall in value.

Just as setting the right policy is crucial, avoiding hasty decisions is likewise important. After China restricted access to rare earths in 2010, Tokyo quickly made the strategic decision to reduce the country's reliance on China's rare earths, encouraging companies not to use materials from China. The Japanese looked prescient as the rare earth price skyrocketed through 2011. But the price quickly crashed because the spike was more related to geopolitics than to economics. And Japan's plan began to backfire.

After an initial rush to stock up on rare earth elements, Japanese companies reduced their use of them dramatically and quickly, especially the light rare earths, the ones Japan used in copious amounts. In the following two years, when prices dropped, they likely fell far lower than they would have due to the dramatic reduction in Japan's demand. The lower prices left many Western rare earth companies struggling. Many of those companies may not last, which would, paradoxically, make Japan more reliant on China and set up long-term dependencies.

Tokyo's hasty decision to help companies to shift from using rare earth elements also had another, more subtle effect. When companies followed government policies and reduced the amount of rare earths in their products, it led to less energy-efficient motors for products ranging from air conditioners to elevators. Although the loss in efficacy from switching to a less efficient magnet or different system may be small for each product, its effect is large when the world is buying, for example, air conditioners that are even just a few percent less efficient.[33]

Governments need to let companies deal with the short-term vagaries of the market. The truth is that unless governments are willing to financially guarantee the development of an international supply chain of raw materials, and to continuously assess it, there is no way to ensure consistent supply. The only time governments should step in is to offer a short-term solution to avoid an economic collapse or to ensure that there are no gaps in a country's military defense supply lines.

As we have seen, countries can only do so much individually to improve resource security because supply lines are global. What's more, while rare metal supply lines are unique, many of the strategies to ensure their viability are similar to the strategies for ensuring supply lines of other natural resources. For example, individual resource-dependent countries need to bolster political ties with neighbors, especially resource-rich ones, as the United States did with Latin America in the first half of the twentieth century. Setting up trade missions and signing trade agreements to increase trade and investment in resources benefits the home country, but the additional production that would result will also benefit the world. Ultimately, however, just as one country cannot dig itself to resource security the world cannot do so either. There is a role for working together.

To avoid future conflict, it would be useful to address rare metal supply concerns in a global forum. But since none exists, it's time to create one. After oil shortages in the early 1970s, sixteen countries founded the International Energy Agency (IEA) to ensure the uninterrupted flow of oil. The agency's mission expanded to "promote diversity, efficiency and flexibility within all energy sectors."[34] We now need an International Materials Agency, an IEA for mineral resources, including rare metals.

It is critical to have an organization that collects statistics, writes market reports, and provides a forum for countries to discuss issues related to natural resources. An International Materials Agency with its own staff can develop strategies for best practices in managing resources. Dialogue and an attempt to make markets transparent is the best hope we have of preventing future resource conflicts. And a materials agency is a far more effective venue than the World Trade Organization for hashing out disagreements, especially for the rare earth materials that go into Masato Sagawa's magnet.

Rare earth magnet manufacturers now use lower amounts of rare earths, including dysprosium, in their magnets than in 2010 when China cut exports to Japan. Part of the reason is that companies, including the one Sagawa runs, have become better at using dysprosium selectively. Other companies have found that adding terbium, a rare earth element produced only in China, can help reduce the need for dysprosium.[35]

When I ask Sagawa about the switch from dysprosium to terbium, he warns me, "The scarcity of terbium will become much more serious than for dysprosium."[36] He adds that using terbium, which is four times less abundant than dysprosium, is a temporary move. But adding terbium reminds me of a similar decision thirty years ago when he added dysprosium to his magnet recipe. Companies may be able to reduce the amount of terbium and dysprosium in magnets by 20–30 percent over a five-year period, but the growth rate of magnets over the same period may well exceed 50 percent or more. This means that despite using these rare earth elements in smaller quantities per magnet, our total demand for rare earths will rise.

We are now at a critical moment. The speed of technological change will soon outpace the ability of our supply lines to produce rare metals at the prices demanded. Meeting this growing demand for rare metals requires profound changes in how we use and sell our products. My fear is that a lack of attention to and understanding of this new dynamic in the Rare Metal Age, as well as a lack of attention and understanding in regard to these critical materials, will limit our prosperity and undermine our environment. My hope is that this book in some measure will serve as a rallying call to inspire a new generation to learn more about the ingredients of our gadgets, guns, and sustainable future.

Notes

1

Metals, Metals Everywhere

1. David Lieberman, "CEO Forum: Microsoft's Ballmer Having a 'Great Time,'" *USA Today*, April 29, 2007, http://usatoday30.usatoday.com/money/companies/management/2007-04-29-ballmer-ceo-forum-usat_N.htm; Jacqui Cheng, "The Truth about the iPhone's Sales Numbers," *Ars Technica*, January 2008, http://arstechnica.com/apple/2008/01/the-truth-about-the-iphones-sales-numbers/; Connie Guglielmo, "Apple iPhone Sold Out at Most Stores after Four Days (Update3)—Bloomberg," July 3, 2007, http://www.bloomberg.com/apps/news?pid=newsarchive&sid=a7A4BDWusr2U; Statista, "Apple iPhone: Global Sales 2007–2015, by Quarter," 2015, http://www.statista.com/statistics/263401/global-apple-iphone-sales-since-3rd-quarter-2007/.

2. Paul Kedrosky, "The Jesus Phone," *Wall Street Journal*, June 29, 2007, http://www.wsj.com/articles/SB118308453151652551; Brian Lam, "The Pope Says Worship Not False Idols: Save Us, Oh True Jesus Phone," *Gizmodo*, December 26, 2006, http://gizmodo.com/224143/the-pope-says-worship-not-false-idols-save-us-oh-true-jesus-phone.

3. Fred Vogelstein, "And Then Steve Said, 'Let There Be an iPhone,'" *New York Times*, October 6, 2013, http://www.nytimes.com/2013/10/06/magazine/and-then-steve-said-let-there-be-an-iphone.html?pagewanted=all; Steve Jobs, "Steve Jobs: Complete Transcript of Steve Jobs, Macworld Conference and Expo, January 9, 2007," *Genius*, January 9, 2007, http://genius.com/Steve-jobs-complete-transcript-of-steve-jobs-macworld-conference-and-expo-january-9-2007-annotated/.

4. Helen Walters, "A Sputnik Moment for STEM Education: Ainissa Ramirez at TED2012," TED Blog, March 2, 2012, accessed November 2, 2014,

http://blog.ted.com/2012/03/02/a-sputnik-moment-for-stem-education
-ainissa-ramirez-at-ted2012/.

5. T. E. Graedel, E. M. Harper, N. T. Nassar, and B. K. Reck, "On the Materials Basis of Modern Society," in *Proceedings of the National Academy of Sciences,* ed. William C. Clark, http://www.pnas.org/content/early/2013 /11/27/1312752110.full.pdf, December 2, 2013, doi:10.1073/pnas.1312752110; Allied Market Research, "Permanent Magnet Motor Market Is Expected To Reach $45.3 Billion, Global, By 2020," January 7, 2015, accessed April 7, 2015, http://www.alliedmarketresearch.com/press-release/permanent-magnet -motor-market-is-expected-to-reach-45-3-billion-global-by-2020-allied -market-research.html; Metal-Pages, "Rapid Growth in Permanent Magnet Market, Worth $18 Billion by 2018," September 2, 2013, accessed December 7, 2014, http://www.metal-pages.com/news/story/73460/rapid-growth -in-permanent-magnet-market-worth-18-billion-by-2018.

6. "At BASF we don't make a lot of the products you buy. We make a lot of the products you buy better." The German chemical manufacture's slogan of the late 1990s is an apt way to describe many of the minor metals. See https://www.youtube.com/watch?v=ZJHPpsb3FzM.

7. Midwest Railcar Corp., "Gondola—100 Ton-52' 6'," accessed April 1, 2015, http://www.midwestrailcar.com/equipGondola100-525.html; Copper Development Association, *Annual Data 2014 Copper Supply & Consumption: 1993–2013,* 1st ed., e-book (New York, 2014), available at http://www .copper.org/resources/market_data/pdfs/annual_data.pdf; Periodic Table, "What Is Iron?" accessed December 26, 2014, http://www.periodic-table .org.uk/element-iron.htm. The problem with the term "minor metal" is that it lacks a purpose. If "minor metal" really describes production levels, it should be called "rare metal" or the more direct "limited production metal." "Minor metal" describes what a metal is not rather than what it is, as in, "it is not a base metal."

8. R. Eggert, "Strategic and Critical Minerals Policy: Domestic Minerals Supplies and Demands in a Time of Foreign Supply Disruptions," testimony before the Subcommittee on Energy and Mineral Resources, Committee on Natural Resources, U.S. House of Representatives, Washington, DC, May 24, 2011. Author's estimates.

9. U.S. Geological Survey, "Byproduct Metals and Rare-Earth Elements Used in the Production of Light-Emitting Diodes—Overview of Principal Sources of Supply and Material Requirements for Selected Markets: U.S. Geological Survey Scientific Investigations Report 2012–5215" (Reston, VA, 2012), available at http://pubs.usgs.gov/sir/2012/5215/.

10. Walter Isaacson, *Steve Jobs* (New York: Simon and Schuster, 2011).

11. C. Hagelüken, R. Drielsmann, and K. Ven den Broeck, "Availability of Metals and Materials," in *Precious Materials Handbook*, Ulla Sehrt and Matthias Grehl, 10–35 (Hanau-Wolfgang, Germany: Umicore AG, 2012).

12. Michael Wolff, "Michael Wolff: Uber Invades the World," *USA-Today*, June 14, 2014, http://www.usatoday.com/story/money/columnist/wolff/2014/06/14/the-rise-of-uber/10417655/; Brian Lam, "The Life of Steve Jobs," *Gizmodo*, August 24, 2011, http://gizmodo.com/5301470/the-life-of-steve-jobs---so-far.

13. Michael Feroli, "Economics Web Note," Morgan Markets, J.P. Morgan, September 10, 2014, https://mm.jpmorgan.com/EmailPubServlet?doc=GPS-938711-0.html&h=-825pgod; Mian Ridge, "iPhone 6 to Boost Taiwan's GDP," Beyondbrics Blog, September 2, 2014, http://blogs.ft.com/beyond-brics/2014/09/02/iphone-6-to-boost-taiwans-gdp; Apple, Form 10-K, October 31, 2012, http://files.shareholder.com/downloads/AAPL/3721939736x0x756594/71AB3488-9710-44D4-B161-197F330FC39A/SEC-AAPL-1193125-12-444068.pdf; *Financial Times* Global 500 companies in 2007, accessed January 10, 2015, www.ft.com/cms/s/0/ad53b3b2-2586-11dc-b338-000b5df10621.html; *Financial Times* Global 500 companies in 2012, accessed January 10, 2015, http://im.ft-static.com/content/images/a81f853e-ca80-11e1-89f8-00144feabdc0.pdf.

14. Jared Diamond, "What's Your Consumption Factor?" *New York Times*, January 2, 2008, http://www.nytimes.com/2008/01/02/opinion/02diamond.html?pagewanted=all.

15. Nomura, Steel Consumption Per Capita (Image), May 2012, http://av.r.ftdata.co.uk/files/2012/05/steel_percapita_china_ROW_nomura.gif; Magnus Ericsson, *Trends in Metals and Mining*, 1st ed., e-book (London: International Council on Metals and Mining, 2012), available at http://www.icmm.com/document/4441; Bert Hofman, "How Urbanization Can Help the Poor," *China Daily*, April 17, 2014, http://www.chinadaily.com.cn/cndy/2014-04/17/content_17439759.htm.

16. In 2012, Alex King, head of the U.S. Critical Materials Institute said, "This is the first time in forty years where I have concerns about the supply of certain materials to meet global demand." Alex King, "Welcome Remarks," Second International Workshop on Rare Metals, Boston, November 25, 2012; American Chemical Society, "Endangered Elements," accessed on April 7, 2015, http://www.acs.org/content/acs/en/greenchemistry/research-innovation/research-topics/endangered-elements.html; John W. Poston Sr., "Do Transuranic Elements Such as Plutonium Ever Occur Naturally?" *Scientific American*, March 23, 1998, http://www.scientificamerican.com/article/do-transuranic-elements-s.

17. Aljazeera, "Israel Unveils Maritime Version of Iron Dome," October 31, 2014, http://www.aljazeera.com/news/middleeast/2014/10/israel -unveils-maritime-version-iron-dome-20141031114213369908.html.

18. Alessandro Bruno, "Mark Smith to Turn NioCorp into One of the Top Niobium Producers in the World," *InvestorIntel*, June 23, 2014, http:// investorintel.com/rare-earth-intel/mark-smith-turn-niocorp-one-top -niobium-producers-world/.

19. But the status quo is difficult to change. As Thomas Graedel, a preeminent professor of material science at Yale University, notes, "you really want a long-term view of material supply, and we have no way to get that information." Thomas Graedel, interview by David Abraham, New Haven, CT, April 8, 2013. Dell, "Addressing Conflict Minerals," accessed April 11, 2015, http://www.dell.com/learn/us/en/uscorp1/conflict-minerals; Jeremy Hsu, "Gadget Makers Face Shortages of Essential Elements," *Tech-NewsDaily, InvestorIntel*, March 10, 2011, accessed November 2, 2014, http:// www.technewsdaily.com/4911-gadgets-cost-fortune-precious-110216html.

20. Thomas Fuller, "Mobile Deal in Myanmar Elicits Anger over Religion," *New York Times*, June 28, 2013, http://www.nytimes.com/2013/06 /28/world/asia/mobile-deal-in-myanmar-elicits-anger-over-religion.html ?ref=global-home; World Bank, "World Development Indicators," 2014, http://wdi.worldbank.org/table/2.8.

2

National Struggles

1. Alan Cowell, "Zaire's Bloody Past Makes Cobalt's Future Uncertain," *New York Times*, August 30, 1981, www.nytimes.com/1981/08/30 /weekinreview/zaire-s-bloody-past-makes-cobalt-s-future-uncertain .html; Congressional Budget Office (CBO), *Cobalt: Policy Options for a Strategic Mineral* (Washington DC, 1982), available at https://www.cbo.gov /sites/default/files/cbofiles/ftpdocs/51xx/doc5126/doc29-entire.pdf.

2. Bohumil Volesky, *Biosorption of Heavy Metals* (Boca Raton, FL: CRC Press, 1990). "[The Soviets] had a shrewd inkling of what was going to blow," Luf Lubett, the former chairman of the Minor Metals Traders Association, commented at the time. Bernard D. Nossiter, "Soviets Reportedly Bought Up Cobalt before Zaire Invasion," *Washington Post*, May 24, 1978; Gary Thatcher, "South Africa How Vital to the West?" *Christian Science Monitor*, December 16, 1980, www.csmonitor.com/1980/1216/121627.html /%28page%29/3; Robert D. Hershey Jr., "U.S. Weighs Subsidizing of Strate-

gic Minerals," *New York Times,* June 13, 1981, www.nytimes.com/1981/06/13 /business/us-weighs-subsidizing-of-strategic-minerals.html. Rep. Jim Santini (D-NV) commented, "The Soviet Union has moved into the international resource arena armed with a strategy of confrontation that extends beyond economic competition, but which falls just short of conventional military conflict." Thatcher, "South Africa How Vital to the West?"

3. CBO, *Cobalt: Policy Options for a Strategic Mineral;* U.S. Congress, Office of Technology Assessment, *Strategic Materials: Technologies to Reduce U.S. Import Vulnerability* (Washington, DC, 1985), OTA-ITE-248, available at www.princeton.edu/~ota/disk2/1985/8525/8525.PDF.

4. "High Anisotropy Field Layer Production Process," *Tech Journal,* www.tdk.co.jp/techjournal_e/vol08_hal/contents02.htm.

5. In the early 1980s, the U.S. Navy and General Motors were also seeking new alternatives for cobalt in magnets. Jan Herbst, "Innovation via Insurrection: Nd-Fe-B Permanent Magnets," paper presented at the fiftieth Conference on Magnetism and Magnetic Materials, San Jose, CA, October 30–November 5, 2005.

6. Masato Sagawa, interview by David Abraham, Kyoto, Japan, July 20, 2013; C. Claiborne Ray, "Q&A: Magnetic Metals," *New York Times,* May 16, 2006, www.nytimes.com/2006/05/16/science/16qna.html. Magnets are made only from iron, nickel, and cobalt due to their atomic structure. In all metals, the spin and rotation of electrons around the atoms of materials produce a tiny electronic charge. In most atoms, electrons come in pairs that spin in opposite directions so the charges cancel out. However, the atoms of iron, cobalt, and nickel have unpaired electrons creating the opportunity for magnetism if these atoms line up.

7. Sagawa, interview, July 20, 2013; Masato Sagawa, "Japan Prize Commemorative Lectures," Japan Prize Foundation, 2012, http://www .japanprize.jp/data/prize/commemorative_lec_2012_e.pdf.

8. Nicola Jones, "Materials Science: The Pull of Stronger Magnets," *Nature* 472, no. 7341 (2011): 22–23, doi:10.1038/472022a, available at www.nature.com/news/2011/110405/full/472022a.html?WT.ec_id =NATURE-20110407. The atomic electrons in the magnet resemble an army brigade. When they are all aligned and facing the same direction with appropriate spacing, they are a strong unit. But when the atoms are not aligned, the magnetism is weak.

9. Stan Trout, telephone interview by David Abraham, November 20, 2014. The U.S. Navy, General Motors, and General Electric were all pursuing similar research.

10. Sagawa, "Japan Prize Commemorative Lectures."

11. Sagawa, interview, July 20, 2013.

12. Andrew H. Malcolm, "Japanese-Chinese Dispute on Isles Threatens to Delay Peace Treaty," *New York Times,* April 15, 1978.

13. Joseph Ferguson, "The Diaoyutai-Senkaku Islands Dispute Reawakened," *Jamestown Foundation,* China Brief, Volume 4, Issue 3, February 4, 2004, www.jamestown.org/single/?no_cache=1&tx_ttnews [tt_news]=3623 .#.VITF71fF9HF; Austin Ramzy, "Japan Releases Chinese Captain, but Tensions Remain," *Time,* September 27, 2010, content.time.com/time/world /article/0,8599,2021625,00.html.

14. Wei Tian, "Arrest Brings Calamity to Trawler Captain's Family," *China Daily,* September 13, 2010, usa.chinadaily.com.cn/2010-09 /13/content_11293724.htm; Yomiuri Shimbun, "Video Shows Clear Hits on JCG Boats," *Asia One News,* November 6, 2010, http://news.asiaone.com /News/Latest%2BNews/Asia/Story/A1Story20101106-245993.html; *China Secret Ship Tries to Ram Japan Coast Guard Ship,* YouTube Video, Hamilton's Military Channel, March 19, 2013, https://www.youtube.com/watch?v =65ozHMEWwmc; sengoku38 コピー転載 流出尖閣衝突ビデオ4; YouTube, 2010. "Sengoku38 コピー転載 流出尖閣衝突ビデオ4" [Copy of the Senkaku collision video], November 4, 2010, accessed December 7, 2014, https:// www.youtube.com/watch?v=YKDG2_osjBs.

15. "Japan Says Activists Must Go to China," *Taipei Times,* March 26, 2004, http://www.taipeitimes.com/News/world/archives/2004/03/26/2003 107798; Andrew H. Malcolm, "Japanese-Chinese Dispute on Isles Threatens to Delay Peace Treaty," *New York Times,* April 15, 1978; Yuki Tatsumi, "Senkau/East China Sea Disputes: A Japanese Perspective," CNA Maritime Asia Project, no. 8-2013, accessed December 7, 2014, http://www.stimson .org/images/uploads/research-pdfs/Yuki-CNA_paper_8-2013.pdf.

16. Roger W. Bowen and Joel J. Kassiola, *Japan's Dysfunctional Democracy: The Liberal Democratic Party and Structural Corruption* (Armonk, NY: M. E. Sharpe, 2003).

17. Kyung Lah, "China Arrests 4 Japanese against Backdrop of Diplomatic Battle," CNN, September 24, 2010, http://www.cnn.com/2010 /WORLD/asiapcf/09/24/china.japanese.arrests/.

18. Keith Bradsher, "Trade Officials Ponder Response to China's Rare Earth Stance," *New York Times,* October 14, 2010, www.nytimes.com/2010 /10/14/business/global/14rare.html. In 2011, the British Geological Survey reported that China produced more than twenty-eight critical materials that the world needed; in 2012 the survey calculated differently, reducing that number to twenty-two along with the amount of total materials deemed

critical. "British Geological Survey: Risk List 2011," British Geological Survey. https://www.bgs.ac.uk/downloads/start.cfm?id=2063.

19. Martin Fackler, "Japan Retreats with Release of Chinese Boat Captain," *New York Times,* September 25, 2010, http://www.nytimes.com/2010/09/25/world/asia/25chinajapan.html?pagewanted=all; General Electric, "Rare Earth FAQs," accessed December 7, 2014, http://www.gelighting.com/LightingWeb/na/images/GE-Rare-Earth-Materials-FAQ.pdf; Sagawa, interview, July 20, 2013.

20. Arnold Magnetic Technologies Corporation, accessed January 1, 2015, http://www.arnoldmagnetics.com/Content1.aspx?id=4627; "How Do Speakers Work?" Explore—Physics.Org, accessed January 1, 2015, http://www.physics.org/article-questions.asp?id=54.

21. EWI, "Rare Earth Materials: Important Industrial Application and Uses," January 2, 2013, accessed December 7, 2014, http://ewi.org/eto/wp-content/uploads/2013/01/2-Major-Industrial-Uses-S.pdf. In 2013, MarketsandMarkets predicted permanent magnets will be worth $18 billion by 2018, growing by 8.7 percent annually. Two years later in 2015, Allied Market Research predicted the market would grow to over $45 billion by 2020. Allied Market Research, "Permanent Magnet Motor Market Is Expected to Reach $45.3 Billion, Global, By 2020," January 7, 2015, accessed April 7, 2015, http://www.alliedmarketresearch.com/press-release/permanent-magnet-motor-market-is-expected-to-reach-45-3-billion-global-by-2020-allied-market-research.html; Metal-Pages, "Rapid Growth In Permanent Magnet Market, Worth $18 Billion by 2018," September 2, 2013, accessed December 7, 2014, http://www.metal-pages.com/news/story/73460/rapid-growth-in-permanent-magnet-market-worth-18-billion-by-2018; Sagawa, interview, July 20, 2013.

22. Karl A. Gschneidner Jr., "Neodymium: Supply, Demand, Substitution and Recycling," Critical Materials Flow in an Age of Constraint, U.S. DOE Office of Intelligence, Woodrow Wilson International Center for Scholars, 2011, accessed December 7, 2014, http://www.wilsoncenter.org/sites/default/files/gschneidner_neodynium_supply_demand_substitution_and_recycling.pdf.

23. G. M. Mudd, Z. Weng, S. M. Jowitt, I. D. Turnbull, and T. E. Graedel, "Quantifying the Recoverable Resources of By-Product Metals: The Case of Cobalt," *Ore Geology Reviews* 55 (2013): 87–98, doi:10.1016/j.oregeorev.2013.04.010.

24. "Made in China," Dan Rather Reports, episode 213, transcript, http://bit.ly/1vtWCJD.

25. Alfred E. Eckes, *The United States and the Global Struggle for Minerals*, (Austin: University of Texas Press, 1979); "Smugglers Go to Prison: One Year and $1,000 Fine for Secreting Tungsten on Swedish Ship," *New York Times*, November 10, 1917, accessed December 7, 2014, http://times machine.nytimes.com/timesmachine/1917/11/10/102647354.html?page Number=4.

26. John Pike, "DLA Strategic Materials," *Globalsecurity*.org, accessed December 7, 2014, http://www.globalsecurity.org/military/agency /dod/dnsc. Eckes, *The United States and the Global Struggle for Minerals*.

27. "Says US Lacks Vital Minerals for War Use," *Sunday Morning Star*, May 13, 1939, http://news.google.com/newspapers?nid=2293&dat =19390514&id=Y84mAAAAIBAJ&sjid=bwIGAAAAIBAJ&pg=1129,27 97807.

28. Peter Howlett, "Economic Disasters of the Twentieth Century," *Economic History Review*, ed. Michael J. Oliver and Derek H. Aldcroft 61, no. 2 (2008), doi:10.1111/j.1468-0289.2008.00432_32.x. Roosevelt had previously announced that a moral embargo against Italy was unsuccessful because American companies increased trade with Italy. David Cortright, *Peace: A History of Movements and Ideas* (Cambridge: Cambridge University Press, 2008); Bertram Hulen, "Moral Embargo of Soviet by U.S.; President Assails 'Obviously Guilty'—Lie to Molotoff Passed on Bombings," December 2, 1939, accessed December 7, 2014, http://query.nytimes .com/mem/archive-free/pdf?res=9407E0DA123EE23ABC4B53DFB46783 82629ED; Robert Dallek, *Franklin D. Roosevelt and American Foreign Policy, 1932–1945* (New York: Oxford University Press, 1995); R. H. Limbaugh, *Tungsten in Peace and War, 1918–1946* (Reno: University of Nevada Press, 2010), 153; "Hyperwar: U.S. Government Manual—1945," accessed December 7, 2014, http://www.ibiblio.org/hyperwar/ATO/USGM/EWA.html. Roosevelt established the U.S. Commercial Company to buy rare metals from nonaligned countries such as Turkey, Spain, and Argentina. J. B. DeMille, *Strategic Minerals: A Summary of Uses, World Output Stockpiles, Procurement* (New York: McGraw-Hill, 1947); http://archive.org/stream/strate gicmineral031804mbp/strategicmineral031804mbp_djvu.txt; J. S. McGrath, "International Aspects of War Mineral Procurement," in *Minerals Yearbook 1942*, ed. E. W. Pehrson, 25–34 (New York and London: McGraw Hill, 1943); http://digicoll.library.wisc.edu/cgi-bin/EcoNatRes /EcoNatRes-idx?type=turn&entity=EcoNatRes.MinYB1942.p0036&id =EcoNatRes.MinYB1942&isize=text; L. Caruana and H. Rockoff, "A Wolfram in Sheep's Clothing: Economic Warfare in Spain, 1940–1944," *Journal of Economic History* 63, no. 1 (March 2003): 100–126. In Europe, the

United States used oil exports as leverage to prevent Spain and Portugal from selling tungsten to Germany; the British ran naval blockades to prevent minerals from reaching Germany; and Germany had U-boats helping shipments of tungsten to reach its shores. The United States and Britain also engaged in bidding wars to keep Iberian tungsten out of the hands of Germany. The price of tungsten jumped from $1,100 in 1940 to $20,000 one year later. Jerry W. Markham, *A Financial History of the United States* (Armonk, NY: M. E. Sharpe, 2002); Inquiry into the Operation of the Reconstruction Finance Corporation and Its Subsidiaries Under Senate Resolution 132, Part 2, hearings before the United States Senate Committee on Banking and Currency, Special Subcommittee to Investigate the Reconstruction Finance Corporation, Eightieth Congress, second session, January 14–16, 22, 1948, p. 1136; "Reich has Plenty of War Minerals," *Montreal Gazette,* January 19, 1942; Associated Press, "Further Cut in Nazi Steel Output Seen as Chrome, Tungsten, etc., Supplies Cut," *Montreal Gazette,* May 8, 1944, http://bit.ly /1wQrTOl; Robert Hershey Jr., "U.S. Weighs Subsidizing of Strategic Minerals," *New York Times,* June 13, 1981, www.nytimes.com/1981/06/13/business /us-weighs-subsidizing-of-strategic-minerals.html.

29. The committee was to study "the accessibility of critical raw materials to the US during a time of war" and "recommend methods of encouraging developments to assure the availability of supplies of such critical raw materials adequate for the expanding economy and the security of the United States." Government Printing Office, *Stockpile and Accessibility of Strategic and Critical Materials to the United States in Time of War* (Washington, DC, 1953); National Archives of the United States, *The Code of Federal Regulations of the United States Title 32A, National Defense Appendix* (Washington, DC: Government Printing Office, 1953).

30. In a letter to Congress in 1982, Reagan's budget director stated that the government would rely on the marketplace to meet the resource needs of the country, despite requests from Democratic and Republican members for the Administration to support domestic production of needed resources. United Press International, "Reagan Rejects Mineral Subsidies," *Florence Times Tri City Daily,* February 1, 1982. Even when Congressional leaders released a four-hundred-page strategic materials report in 1985, aimed at reducing U.S. import vulnerabilities on Soviet and African rare metals, the administration's predilection was to let the market work. *Strategic Materials: Technologies to Reduce U.S. Import Vulnerability* (Washington, DC: Congress of the U.S. Office of Technology Assessment, 1985), available at https://www.princeton.edu/~ota/disk2/1985/8525/8525 .PDF.

31. The Defense Logistics Agency sold off $5.9 billion worth of the approximately $7.4 billion total of these metals in the 1990s. Defense Logistics Agency, accessed December 14, 2014, www.dla.mil/Pages/default _old2.aspx; United Nations Environment Programme, *Decoupling Natural Resource Use and Environmental Impacts from Economic Growth*, 2011, http://www.unep.org/resourcepanel/decoupling/files/pdf/decoupling _report_english.pdf.

32. Tao Xu, *The Development and Restructuring of China's Metal Mining Industry* (Kingston, ON: Centre for Resource Studies, Queen's University, 1991).

33. Liu Chang, "Changing Nonferrous Market, Nonferrous Metal Review," in Xu, *The Development and Restructuring of China's Metal Mining Industry*. China has a history of producing materials cheaply after World War I; it produced tungsten at a quarter of the U.S. price and sold it abroad, angering miners in the United States as the market collapsed; Limbaugh, *Tungsten in Peace and War*; Tatsuo Ota, "Rare Earth Resources and Related Industries in Japan," *Journal of MMIJ* 127, no. 9 (2011): 549; Antony B. T. Werner and W. D. Sinclair, *International Strategic Mineral Issues Summary Report—Tungsten* (Washington: General Printing Office, 1998), accessed December 17, 2014, available at http://pubs.usgs.gov/pdf/circular /pdf/c930-o.pdf; David O'Brock, interview by David Abraham, Sillamäe, Estonia, January 24, 2013.

34. "China Rolls Over Annual Export Quotas for Antimony, Tungsten, Molybdenum and Indium," Metal-Pages, November 11, 2013, http:// www.metal-pages.com/news/story/75124/china-rolls-over-annual-export -quotas-for-antimony-tungsten-molybdenum-and-indium/. The Chinese export quotas for minor metals were unchanged in 2014. "The export quota for indium is 231 tonnes, while the quota for molybdenum is 25,000 tonnes. Beijing also set the annual quota for antimony and tungsten at 59,400 tonnes and 15,400 tonnes, respectively;" Xu, *The Development and Restructuring of China's Metal Mining Industry*. Gan Yong, "Opening Speech," at the seventh International Conference on Rare Earth Development and Application and the second China Rare Earth Summit, Ganzhou, China, August 11, 2013.

35. Karl Gerald van den Boogaart, Polina Klossek, and Andreas Klossek, "How Forward Integration along the Rare Earth Value Chain Threatens the Global Economy," at the Society for Mining, Metallurgy and Exploration's Critical Minerals 2014 conference, Denver, August 3–5, 2014.

36. Michael Silver, telephone interview by David Abraham, March 13, 2014.

37. "How Minerals Will Impact Global Geopolitics in the 21st Century," live Webcast of American Elements CEO Michael Silver's address to the New York Mineralogical Club, October 9, 2013, accessed December 14, 2014, http://www.americanelements.com/presentations2013.html.

38. Chang-Ran Kim, "China Seeks Japan Tech in Rare Earth Deals: Reports," Reuters, September 7, 2011, accessed December 14, 2014, www.reuters.com/article/2011/09/07/us-china-rareearth-japan-idUSTRE 7860S920110907.

39. Baris Karapinar, "Export Restrictions on Natural Resources: Policy Options and Opportunities for Africa," World Trade Institute, University of Bern, Switzerland, n.d., http://www.wti.org/fileadmin/user_upload/nccr -trade.ch/news/TRAPCA%20Paper%20%28Submitted1711%29_BK.pdf.

40. European Union, Mexico, United States vs. China: "China: Raw Materials," World Trade Organization, 2013, http://www.wto.org/english /tratop_e/dispu_e/cases_e/1pagesum_e/ds398sum_e.pdf; European Union, "China: Measures Related to the Exportation of Rare Earths, Tungsten and Molybdenum," World Trade Organization Panel Proceedings, October 30, 2012, http://trade.ec.europa.eu/doclib/docs/2013/april/tradoc_150806.pdf; "China: Measures Related to the Exportation of Rare Earths, Tungsten and Molybdenum," World Trade Organization, August 29, 2014, accessed December 14, 2014, www.wto.org/english/tratop_e/dispu_e/cases_e/ds431_e .htm; "How Would a Potential Scrapping of the Chinese Export Quota and Tariffs System Impact the Rare Earths Market?" Roskill Briefing Paper, June 2014, accessed December 14, 2014, http://www.roskill.com/news/how-would -a-potential-scrapping-of-the-chinese-export-quota-and-tariffs-system -impact-the-rare-earths-market.

41. CRU Consulting, "Trends in Capital Expenditure," presentation in Santiago, Chile, April 4, 2011; "Strategic and Critical Materials Report on Stockpile Requirements," Office of the Under Secretary of Defense for Acquisition, Technology and Logistics, January, 2013, http://mineralsmake life.org/assets/images/content/resources/Strategic_and_Critical_Materi als_2013_Report_on_Stockpile_Requirements.pdf.

3

Corporate Hurdles

1. Malapronta, "Tauá Grande Hotel e Termas de Araxá em Araxá, MG," accessed December 27, 2014, http://www.malapronta.com.br/hotel 1469-taua-grande-hotel-e-termas-de-araxa.

2. Araxá town officials, interview by David Abraham, Araxá, Brazil, May 13, 2013.

3. Nicholas Riccardi, "Mining Towns Not All Happy with This Boom," Los Angeles Times, October 1, 2008, http://articles.latimes.com/2008/oct/01/nation/na-mine1.

4. U.S. Energy Information Administration (EIA), "Saudi Arabia," 2014, http://www.eia.gov/countries/country-data.cfm?fips=sa; James F. Carlin Jr., "Antimony" (Reston, VA: U.S. Department of the Interior, 2012); Thomas G. Goonan, "Lithium Use in Batteries, Circular 1371" (Reston, VA: U.S. Department of the Interior, 2014).

5. Wikileaks, "Cable Viewer," February 18, 2009, https://wikileaks.org/cable/2009/02/09STATE15113.html.

6. CBMM, "Certifications," 2014, http://www.cbmm.com/us/p/20/certifications.aspx. CBMM was the first mining company to receive ISO 14001, Environmental Management System Certification; Tadeu Carneiro, interview by David Abraham, Araxá, Brazil, May 14, 2013; Marta Vieira, "CBMM Investe R$ 1 Bilhão Para Fábrica Em Araxá," August 13, 2013, accessed November 20, 2014, http://www.em.com.br/app/noticia/economia/2013/08/13/internas_economia,434415/cbmm-investe-r-1-bilhao-para-fabrica-em-araxa.shtml.

7. National Research Council, U.S. Committee on Engineering and Technical Systems, "Implications for Processing Strategic, and Materials. Basic and Strategic Metals Industries: Threats and Opportunities: Report," (Washington, DC: National Academy Press, 1985).

8. Cristiane Cuadros, "Brazil's Richest Family Forging $13 Billion Niobium Dream," Bloomberg, March 13, 2013, http://www.bloomberg.com/news/2013-03-13/brazil-s-richest-family-forging-13-billion-niobium-dream.html.

9. Reuters, "Chinese Firms Take 15-Pct Stake in Niobium Producer CBMM," September 1, 2011, http://www.reuters.com/article/2011/09/01/cbmm-niobium-idUSL4E7K11GL20110901; Cuadros, "Brazil's Richest Family."

10. Department of Energy, Industrial Technologies Program, "Mining Industry Energy Bandwidth Study," 2007, accessed December 27, 2014, http://energy.gov/sites/prod/files/2013/11/f4/mining_bandwidth.pdf; Abrahão Issa Filho, Bruno F. Riffel, and Clovis A. de Faria Sousa, "Some Aspects of the Mineralogy of CBMM Niobium Deposit and Mining Pyrochlore Ore Processing" (Araxá, Brazil: Companhia Brasileira de Metalurgia e Mineração, 2001), http://www.cbmm.com.br/Repositorio/Media/site/internas/operations/minerologyaspectsniobiumdeposit.pdf.

11. Filho, Riffel, and Sousa, "Some Aspects of the Mineralogy of CBMM."

12. "Eiffel Tower by Gustave Eiffel," accessed December 28, 2014, http://www.greatbuildings.com/buildings/Eiffel_Tower.html; L. Meyer, "History of Niobium as a Microalloying Element," in *Proceedings of the International Symposium Niobium 2001*, 359–77 (Bridgeville, PA: Niobium Science and Technology, 2001), available at http://www.cbmm.com.br /portug/sources/techlib/science_techno/Table_content/sub_3/images /pdfs/025.pdf.

13. "BBC on This Day | 19 | 1954: 'Metal Fatigue' Caused Comet Crashes," accessed December 28, 2014, http://news.bbc.co.uk/onthisday/hi /dates/stories/october/19/newsid_3112000/3112466.stm.

14. Meyer, "History of Niobium."

15. Ibid.

16. Ibid.

17. S. Wilmes and G. Zwick, "Effect of Niobium and Vanadium as an Alloying Element in Tool Steels with High Chromium Content," Sixth International Tooling Conference, 269–87, published proceedings, http://www .kau.se/sites/default/files/Dokument/subpage/2010/02/21_269_287_pdf _18759.pdf; Tadeu Carneiro, "Niobium, Grams Savings Tonnes," presentation, Rio de Janeiro, September 22, 2010, 14th Americas School of Mines, http://www.pwc.com.br/pt/eventos-pwc/school-of-mines/english/assets /plenary/11h30-risks-and-opportunities-tadeu-carneiro.pdf.

18. "CBMM Nurtures Niobium Market," *Metal Bulletin Monthly*, 1990, 49; Carneiro, interview, May 14, 2013.

19. John Muchira, "Kenya's Kwale Embarks on Production This Month after Two Decades of False Starts," *Mining Weekly*, September 27, 2013, http://www.miningweekly.com/article/kenyas-kwale-starts-produc tion-this-months-after-two-decades-of-false-starts-2013-09-27.

20. Tiomin Resources, *Annual Report*, 1997, http://www.claudiaforgas .com/Tiomin.pdf; United Nations Conference on Trade and Development, *Investment Policy Review Kenya* (New York and Geneva: 2005); Jean-Charles Potvin, e-mail, October 8, 2014.

21. *The Nation*, Kenya, "Tiomin Wants Sh200 Million Suit Cost from Farmers," Socio Economic Network (blog), 2007, http://socioeconomicfo rum50.blogspot.com/2013/04/tiomin-secures-land-for-kenya-plant_16.html; MiningWatch Canada, "Kwale Dispatch: Investigating Tiomin Resources' Criminal Activities in Kenya," August 7, 2007, http://www.miningwatch.ca /kwale-dispatch-investigating-tiomin-resources-criminal-activities-kenya; Potvin, e-mail, October 8, 2014.

22. Canadian Press, "Tiomin Wins Ruling on Fight Against Land Acquisition," Resource Investor, December 19, 2006, http://www.resource investor.com/2006/12/19/tiomin-wins-ruling-on-fight-against-land -acquisiti; "Cabinet Formally Approves Tiomin's Kwale Titanium Mining Lease," June 25, 2004, Press release, http://www.thefreelibrary.com/Cabinet +Formally+Approves+Tiomin   percent27s+Kwale+Titanium+Mining+Le ase.-a011`8797376; International Chamber of Commerce, 2014, Africa Review 2003/04: The Economic and Business Report, 25th ed. https://books.google .co.id/books?id=nDSR1FF2hY8C; CNN, "Tiomin Appoints Robert Jackson as President," August 17, 2006, http://money.cnn.com/news/newsfeeds /articles/marketwire/06155037.htm; Potvin, e-mail, October 8, 2014.

23. Reuters, "RPT-Australia's Base Resources Begins Titanium Mining in Kenya," October 13, 2013, http://www.reuters.com/article/2013/10/13 /kenya-titanium-idUSL6N0I30P020131013.

24. Daniel Cassman, "Revolutionary Armed Forces of Colombia: People's Army," Mapping Militant Organizations, 2012, http://www.stan ford.edu/group/mappingmilitants/cgi-bin/groups/view/89; Michael Smith, "How Colombian FARC Terrorists Mining Tungsten Are Linked to Your BMW Sedan," Bloomberg, August 8, 2013, http://www.bloomberg.com/news /2013-08-08/terrorist-tungsten-in-colombia-taints-global-phone-to-car -sales.html; Patrick Stratton and David Henderson, "Tantalum Market Overview MMTA," January 31, 2013, http://www.mmta.co.uk/tantalum -market-overview; Araxá town officials, interview, May 13, 2013; Cam Simpson, "The Deadly Tin inside Your Smartphone," Businessweek, August 23, 2012, http://www.businessweek.com/articles/2012-08-23/the-deadly-tin -inside-your-ipad.

25. Barry Porter and Soraya Permatasari, "Lynas CEO Finds Social Media Hobbles Rare-Earths Plans," Bloomberg, July 1, 2012, http://www .bloomberg.com/news/2012-07-01/lynas-ceo-finds-social-media-hobbles -rare-earths-plans.html; Global Times, "Do Not Foment Youngsters to Protest," July 7, 2012, http://english.peopledaily.com.cn/90780/7868299.html.

26. Neil Irwin, "What Bourban Producers Can Teach the Oil Industry," New York Times, November 18, 2014, http://www.nytimes.com/2014/11 /18/business/what-bourbon-producers-can-teach-the-oil-industry.html.

27. Metal-Pages, "Metal Prices, Rare Earths," 2014, http://www.metal -pages.com/metalprices/rareearths/.

28. IntierraRMG, "Mining Sector Fund-Raising Drops 24%," May 14, 2013, http://intierrarmg.com/Libraries/Brochures_and_Flyers/Mining _sector_fund-raising_drops_24.sflb.ash. Investment in junior mining companies dropped 24 percent. Deloitte, "Tracking the Trends: The Top 10 Is-

sues Mining Companies Will Face in the Coming Year," 2014, https://www2
.deloitte.com/content/dam/Deloitte/global/Documents/Energy-and
-Resources/dttl-er-Tracking-the-trends-2014_EN_final.pdf.

29. The report refers to internationally listed mining companies. In-
tierrarmg.com, "Exploration Report Reveals US$1.49 Billion in Financing
Deals," February 26, 2013, http://www.intierrarmg.com/articles/13-02-26
/Exploration_report_reveals_US_1_49_billion_in_financing_deals.aspx.

30. The company adds, "Most explorers can't realistically raise the
necessary funds to bring themselves to the stage where they can generate
cash flow from metals production. . . . Neither retail nor professional investors have the capacity, or appetite, to fund equity raisings." IntierraRMG,
"Mining Sector Fund-Raising Drops 24%," PwC, Mine: A Confidence Crisis 2013, http://www.pwc.com/en_GX/gx/mining/publications/assets/pwc
-mine-a-confidence-crisis.pdf; Justin Pugsley, "Junior Mining Companies
Embrace Social Media to Reach Small Investors," Metal-Pages the Blog,
July 23, 2012, http://blog.metal-pages.com/2012/07/23/junior-mining-com
panies-embrace-social-media-to-reach-small-investors/.

31. Bill Radvak, "American Vanadium," July 2013, http://www
.americanvanadium.com/cms-assets/documents/125931-627924.american
vanadiumaug2013.pdf; U.S. Geological Survey, Vanadium (Reston, VA: U.S.
Department of the Interior, 2012), http://minerals.usgs.gov/minerals/pubs
/commodity/vanadium/mcs-2012-vanad.pdf.

32. John Emsley, Nature's Building Blocks (Oxford: Oxford University Press, 2011).

33. David Wogan, "Vanadium Flow Batteries Could Become a Cost
Effective Solution for Balancing Texas' Power Grid," Blog, Plugged In, October 21, 2013, http://blogs.scientificamerican.com/plugged-in/2013/10/21
/vanadium-flow-batteries-could-become-a-cost-effective-solution-for
-balancing-texas-power-grid/.

34. Carneiro, interview, May 14, 2013.

35. CRU Strategies, "Trends in Capital Expenditure," presentation,
Santiago, Chile, 2011. Received through e-mail correspondence with Andrew Leyland, November 7, 2011. United Nations Environment Programme,
Decoupling Natural Resource Use and Environmental Impacts from Economic Growth, 2011, http://www.unep.org/resourcepanel/decoupling/files
/pdf/decoupling_report_english.pdf; Ana Komnenic, "Copper Costs Up,
Grades Down: Metals Economics Group," Mining.com, July 23, 2013,
http://www.mining.com/copper-costs-up-grades-down-metals-economics
-group-99686/; Pat Taylor, interview by David Abraham, Golden, Colorado,
July 12, 2013.

36. Avalon Ventures, Ltd., "The Thor Lake Heavy Rare Earth Deposit, Canada: Advancing to Feasibility," investor presentation, September 4, 2008; Brenda Bouw, "Rare Metals Bring Rare Opportunities to Mining," *Globe and Mail*, March 23, 2010, http://www.theglobeandmail.com/report -on-business/small-business/sb-growth/sustainability/rare-metals-bring -rare-opportunities-to-mining/article4259001/; Don Bubar, "The Thor Lake (Nechalacho) Heavy REE Deposit: Advancing to Feasibility," investor presentation, November 18, 2009.

37. C. K. Gupta and N. Krishnamurthy, *Extractive Metallurgy of Rare Earths* (Boca Raton, FL: CRC Press, 2005), 170; Alain Leveque, lecture, Pro-Edge Second Annual Technology Metals, April 21–22, 2013, Toronto.

38. Matt Gowing, *2011 Rare Earth Industry Update: We Remain Bull-ish*, Mackie Research Capital Corporation, http://www.slideshare.net /RareEarthsRareMetals/mackie-research-capital-corporation-rare-earth -industry-update.

39. Jeff Phillips, phone interview by David Abraham, November 8, 2013.

40. Dudley Kingsnorth, interview by David Abraham, Shanghai, China, September 10, 2013.

41. Elsewhere, for example, where the ocean floor descends beneath continental plates such as along the West Coast mountain ranges of North and South America, magmas formed rich mineral deposits full of copper and minor metals including selenium and tellurium and also formed the mountains. Klaus Schulz, telephone interview by David Abraham and follow-up e-mail, November 7, 2013.

42. Avalon Rare Metals, Inc., "Avalon Reports Promising Initial Results from Nechalacho Project Metallurgical Process Optimization Work," August 1, 2013, http://avalonraremetals.mwnewsroom.com/press-releases /avalon-reports-promising-initial-results-from-nech-tsx-avl-2013080108898 00001.

43. John Sykes, telephone interview by David Abraham, April 20, 2014. Some mining truth telling has included selective sampling of the most mineral-rich ores for resource estimates, favorable pricing of minerals for cost estimates, and recovery rates that are overly optimistic based on technology still under development. Graham Lumley, "Mine Planners Lie with Numbers," GBI Mining, 2011, http://www.scribd.com/doc/81005939/White -Paper-Mine-Planners-Lie-With-Numbers#scribd.

44. Ernst & Young, "Effective Mining and Metals Capital Project Execution: The Drivers of Mining and Metals Project Execution Complexity," accessed December 29, 2014, www.ey.com/GL/en/Industries

/Mining---Metals/Effective-mining-and-metals-capital-project
-execution---The-drivers-of-mining-and-metals-project-execution
-complexity; Lumley, "Mine Planners Lie with Numbers"; Tony Ottaviano,
"Iron Ore: Industry Outlook," Presentation, AJM Global Iron Ore and Steel
Forecast Conference, Perth, Australia Journal of Mining Iron Ore and
Steel Forecast Conference, 2011, http://www.bhpbilliton.com/home
/investors/reports/Documents/110322GlobalIronOreAndSteelForecastCo
nference.pdf.

45. Anthony DePalma, "David Walsh, 52, Promoter Caught Up in
Gold Mine Scandal," *New York Times,* June 6, 1998, http://www.nytimes
.com/1998/06/06/business/david-walsh-52-promoter-caught-up-in-gold
-mine-scandal.html; Clyde H. Farnsworth, "Investing IT; Bre-X Has a Cin-
derella Story to Tell," *New York Times,* March 24, 1996, http://www.ny
times.com/1996/03/24/business/investing-it-bre-x-has-a-cinderella-story
-to-tell.html; Howard Schneider, "A Lode of Lies: How Bre-X Fooled Everyone," *Washington Post,* May 18, 1997, http://www.washingtonpost.com
/wp-srv/inatl/longterm/canada/stories/brex051897.htm.

46. Rare Earth Elements, "The Rare Earth Elements: Chemistry and
Applications," accessed December 28, 2014, http://www.rareearthelements
.us/the_17_elements; Tracy Weslosky, "The 'Rarest of the Rare' Thulium,
a Heavy Rare Earth Element," accessed December 28, 2014, http://thulium
facts.com/.

47. BullionVault, "Gold Price Chart: Live Spot Gold Rates and Silver
Prices," 2014, http://www.bullionvault.com/gold-price-chart.do.

48. George Rapp Jr. and Christopher Hill, *Geoarchaeology: The Earth-
Science Approach to Archaeological Interpretation,* 2nd ed. (New Haven, CT:
Yale University Press, 2006).

49. Metal-Pages, "Minor Metals Prices," 2015, http://www.metal
-pages.com/metalprices/minors.

50. Phillips, phone interview, November 8, 2013.

51. Danny Lehrman, interview by David Abraham, New York, November 14, 2013.

52. Avalon Rare Metals, Notice of Meeting, Information Circular,
Management Discussion and Analysis and Consolidated Financial Statements for the Year Ended August 31, 2013. These are public documents available at www.sedar.com.

53. Avalon Rare Metals, General Proxy Information, Management
Information Circular, 2012.

54. NioCorp, "Largest rare earth mine in the world discovered in
Nebraska," Press release, May 6, 2010, http://www.niocorp.com/index.php

/press-releases/media/50-largest-rare-earth-mine-in-the-world
-discovered-in-nebraska. The website also notes: "Elk Creek is the *only* Niobium deposit under development in the U.S."

4

Production Difficulties

1. Sillamäe town officials, interview by David Abraham, Sillamäe, Estonia, January 22, 2013.

2. Sillamäe, "History: Sillamäe Linn," accessed December 17, 2014,
http://www.sillamae.ee/en/web/eng/history.

3. Sillamäe, "Secret Period: Sillamäe Linn," accessed December 18,
2014, http://www.sillamae.ee/en/web/eng/secret-period; Olaf Mertelsmann,
"The Uranium Enrichment Factory in Sillamäe (Kombinat 7)," Estonica,
January 28, 2009, http://www.estonica.org/en/The_Uranium_Enrichment
_Factory_in_Sillam%C3%A4e_Kombinat_7/.

4. "Tech Topics: The Leading Edge," *Kemet*, September, 2001, Vol 11,
No. 1, http://www.kemet.com/Lists/TechnicalArticles/Attachments/143
/V11N01%20Nb%20vs%20Ta%20Capacitors.pdf; David O'Brock, interview
by David Abraham, Sillamäe, Estonia, January 22, 2013; Sally Lowder,
"David Trueman: The Ups and Downs of Minor Metal Investing," *Gold
Report*, July 5, 2011, accessed December 18, 2014, http://www.theaureport
.com/pub/na/david-trueman-the-ups-and-downs-of-minor-metal-investing.

5. C. K. Gupta and N. Krishnamurthy, *Extractive Metallurgy of Rare
Earths* (Boca Raton, FL: CRC Press, 2005), 352; Richard Hammen,
telephone interview by David Abraham, June 10, 2014; Joseph Ogando,
"News: Recovering from a Shattered HIP," Designnews.com, August 25,
2000, http://www.designnews.com/document.asp?doc_id=225930.

6. Gupta and Krishnamurthy, *Extractive Metallurgy of Rare Earths*,
170; Alain Leveque, Lecture, ProEdge Second Annual Technology Metals,
Toronto, April 21–22, 2013.

7. Tasman Metals, "Rare Earth Elements Ores and Minerals," http://
www.tasmanmetals.com/s/OresMinerals.asp; Gupta and Krishnamurthy,
Extractive Metallurgy of Rare Earths, 59.

8. O'Brock, interview, January 22, 2013.

9. American Chemical Society, "Separation of Rare Earth Elements—
American Chemical Society," n.d., accessed April 11, 2015, https://www.acs
.org/content/acs/en/education/whatischemistry/landmarks/earthele
ments.html.

10. Gupta and Krishnamurthy, *Extractive Metallurgy of Rare Earths*, 7.

11. James Hedrick, "Rare Earths," U.S. Geological Survey, 2004, accessed December 18, 2014, http://minerals.usgs.gov/minerals/pubs/commodity/rare_earths/rareemyb04.pdf.

12. Gupta and Krishnamurthy, *Extractive Metallurgy of Rare Earths*, 93.

13. Dwight Bradley, *A Preliminary Deposit Model for Lithium Brines*, U.S. Department of the Interior and U.S. Geological Survey, 2013, http://pubs.usgs.gov/of/2013/1006/OF13-1006.pdf; Melissa Pistilli, "POSCO's Lithium Brine Processing Technology Could Be a Game Changer," *Lithium Investing News*, April 4, 2013, accessed December 18, 2014, lithiuminvestingnews.com/7146/posco-lithium-brine-processing-technology-extraction-li3-energy-simbol-chile-signumbox/; "Super Expensive Metals," Periodic Table of Videos, May 28, 2013, http://youtu.be/Fg2WzCzKpYU. Making platinum takes about six weeks, and rhodium, which is recovered from the same ore, takes an additional fourteen.

14. Andrew Hunt, "Phytocat: Catalysing the Growth in Metal Recovery," paper presented at the 2nd International Workshop on Rare Metals, Materials Research Society, Boston, November 25, 2012.

15. John Sykes, e-mail, November 8, 2013. Information based on Frontier publicly released statements.

16. See http://www.molycorp.com/about-us/our-facilities/molycorp-mountain-pass/; Tony Mariano, telephone interview by David Abraham, November 12, 2013.

17. Keith R. Long et al., *The Principal Rare Earth Elements Deposits of the United States: A Summary of Domestic Deposits and a Global Perspective* (Reston, VA: U.S. Department of the Interior, U.S. Geological Survey, 2010), available at http://pubs.usgs.gov/sir/2010/5220; Luisa Moreno, "Demand Ahead: Metal Pages Rare Earth Conference in Shanghai," Europac Commentary, 2013, http://www.europac.ca/commentaries/demand_ahead_metal_pages_rare_earth_conference_shanghai.

18. Thomas Graedel, interview by David Abraham, New Haven, CT, April 8, 2013; U.S. Geological Survey, "Mineral Commodity Summaries 2013," January 24, 2013, http://minerals.usgs.gov/minerals/pubs/mcs/2013/mcs2013.pdf.

19. AZO Materials, "Tellurium Dioxide (TeO2): Properties and Applications," accessed December 18, 2014, http://www.azom.com/article.aspx?ArticleID=5817; Martin Lokanc, Roderick Eggert, and Michael Redlinger, "The Availability of Indium: The Present, Medium Term, and Long Term,"

Technical Report NREL/SR-6A20-62409, July 2014. This is from a National Renewable Energy Laboratory technical report prepublication version; Laura Talens Peiro et al., "Rare and Critical Metals as By-Products and the Implications for Future Supply," working paper, 2011, http://www.insead .edu/facultyresearch/research/doc.cfm?did=48916.

20. John Peacey, e-mail, March 1, 2014; Martin Lokanc, Roderick Eggert, and Michael Redlinger, "The Availability of Indium: The Present, Medium Term, and Long Term," Technical Report NREL/SR-6A20-62409, July 2014. This is from a National Renewable Energy Laboratory technical report prepublication version. "Molybdenum output does not necessarily follow the demand for molybdenum but the demand for copper." Hans Imgrund and Nicole Kinsman, "Molybdenum: An Extraordinary Metal in High Demand," *Stainless Steel World*, September 2007, http://www.imoa .info/download_files/molybdenum/Molybdenum.pdf.

21. Michael W. George, "Tellurium," U.S. Geological Survey, Mineral Commodity Summaries, February 2014, accessed December 18, 2014, http:// minerals.usgs.gov/minerals/pubs/commodity/selenium/mcs-2014-tellu .pdf; U.S. Geological Survey figures, Metal-Pages prices, author's estimates.

22. Lokanc, Eggert, and Redlinger, "The Availability of Indium," 4.

23. Ibid., 3.

24. Tia Ghose, "New Atom-Smashing Magnet Passes First Tests," *Live Science*, July 12, 2013, http://www.livescience.com/38129-large-hadron -collider-magnet-passes-tests.html; Lynn Yaris, "Successful Test of New U.S. Magnet Puts Large Hadron Collider on Track for Major Upgrade," *News Center,* July 11, 2013, http://newscenter.lbl.gov/2013/07/11/new-magnet-for -lhc/.

25. "Molycorp-Silmet AS, David O'Brock's interview," Ruslan TV, April 16, 2011, http://www.youtube.com/watch?v=JMZMjiCWuJA; O'Brock, interview, January 22, 2013.

26. Gupta and Krishnamurthy, *Extractive Metallurgy of Rare Earths,* 296.

27. T. S. S. Dikshith, *Hazardous Chemicals Safety Management and Global Regulations* (Boca Raton, FL: CRC Press, Taylor and Francis Group, 2013).

28. Clifford N. Brandon III, "Emerging Workforce Trends in the U.S. Mining Industry," Society for Mining, Metallurgy and Exploration, January 3, 2012, accessed December 18, 2014, http://www.smenet.org/docs/public /EmergingWorkforceTrendsinUSMiningIndustry1-3-12.pdf;   Deloitte, "Tracking the Trends 2011: The Top 10 Issues Mining Companies Will Face in the Coming Year," 2010, http://www2.deloitte.com/content/dam/Deloitte

/ca/Documents/international-business/ca-en-ib-tracking-the-trends-2011
.pdf; various Interviews with Avalon employees, 2012, 2014.

29. Mia Boiridy, e-mail, September 23, 2014; Corby Anderson, phone interview by David Abraham, September 2011.

30. Brandon, "Emerging Workforce Trends."

31. Steve Yue, e-mail, December 20, 2013.

32. Valerie Bailey Grasso, "Rare Earth Elements in National Defense: Background, Oversight Issues, and Options for Congress," Congressional Research Service, 2013, http://www.fas.org/sgp/crs/natsec/R41744.pdf.

5

Trading Networks

1. Ministry of Economy, Trade and Industry, "Strategy to Secure Rare Metal," 2009, Tokyo.

2. In Russian, "Doveryai no Proveryai." Wikipedia, "Trust, but Verify," last modified March 25, 2014, en.wikipedia.org/wiki/Trust,_but_verify.

3. Michael Rapaport, telephone interview by David Abraham, January 7, 2014.

4. The London Metal Exchange recently listed cobalt and molybdenum on the exchange, but the performance has been mixed.

5. Chad Bray, "Regulator Fines Barclays Over the Pricing of Gold," New York Times, May 23, 2014, http://dealbook.nytimes.com/2014/05/23/barclays-fined-43-9-million-in-setting-price-of-gold/?hp&_r=0.

6. "Metal Bulletin, 60% of voters rejected MMTA-LME online pricing proposal," November 30, 2009, accessed October 30, 2014, www.metal bulletin.com/Article/2350149/60-of-voters-rejected-MMTA-LME-online-pricing-proposal.html.

7. Nigel Tunna, interview by David Abraham, Ganzhou, China, August 11, 2013.

8. In 1992, several Canada-based telemarketing companies sold indium directly to investors at inflated prices before going out of business several years later, after law-enforcement investigations in the United States and Canada. Robert D. Brown Jr., "Indium," in Minerals Yearbook, Vol. 1, Metals and Minerals (Washington, DC: U.S. Bureau of Mines and U.S. Department of the Interior, 1996), accessed October 30, 2014, http://minerals.usgs.gov/minerals/pubs/commodity/indium/490494.pdf.

9. Kotaro Itsuki, "China's Rare-Earths Exchange Feeding Price Spikes," Nikkei Asian Review, October 2, 2014, asia.nikkei.com/Markets

/Commodities/China-s-rare-earths-exchange-feeding-price-spikes; Amy C. Tolcin, "Indium," in Mineral Commodity Summaries (2014): 74–75, accessed December 8, 2014, http://minerals.usgs.gov/minerals/pubs /commodity/indium/mcs-2014-indiu.pdf.

10. "Indium and the Investment Industry: Strategic Metal Report," May 13, 2013, Strategic-metal.typepad.com; Metal-Pages, "Analysis: Fanya Metal Exchange Changes Character of Indium Market, Other Metals Could Follow," www.metal-pages.com, December 30, 2013, accessed October 30, 2014, www.metal-pages.com/news/story/76200/analysis-fanya-metal-ex change-changes-character-of-indium-market-other-metals-could-follow. Fanya market participants trade lower-grade indium that would be challenging for refiners to turn into the type of indium needed for flat screens and smartphones.

11. Metal-Pages, "Interview: Fanya Metals Exchange—An Introduction and Its Relation to the Global Market," September 17, 2013, accessed October 30, 2014, www.metal-pages.com/news/story/73762/interview-fanya -metals-exchange—an-introduction-and-its-relation-to-the-global-market.

12. James T. Areddy, "China Nurtures Futures Markets in Bid to Sway Commodity Prices," Wall Street Journal, October 12, 2009, accessed October 30, 2014, online.wsj.com/news/articles/SB125529874012778991.

13. Metal-Pages, "No Impact from Cut in China Indium Duty, but It May Point to Future Intentions," December 17, 2013, accessed October 30, 2014, www.metal-pages.com/news/story/76011/no-impact-from-cut-in -china-indium-duty-but-it-may-point-to-future-intentions.

14. Neil Gough, "Lawsuit Adds to Concern over China Commodities Fraud," New York Times, June 27, 2014, accessed October 30, 2014, deal-book.nytimes.com/2014/06/27/lawsuit-adds-to-concern-over-china -commodities-fraud; Chanyaporn Chanjaroen, "ABN Amro Unmoved by Qingdao Boosting China Commodity Finance," Bloomberg News, September 29, 2014, accessed December 1, 2014, www.bloomberg.com/news/2014 -09-29/abn-amro-commits-to-china-commodity-finance-growth-post -qingdao.html.

15. Anonymous trader, interview by David Abraham, April 14, 2014.

16. Victor L. Hou, "Derivatives and Dialectics: The Evolution of the Chinese Futures Market," New York Law Review 72 (1997): 175; "Derivatives Exchanges: The Mung Bean and Its Adventures," Economist, May 27, 1995, 70; Joseph Kahn, "China Ends Clean-Up Campaign by Licensing Commodity Markets," Asian Wall Street Journal, October 27, 1994.

17. Anonymous trader, e-mail, December 8, 2014; Zhu Sanzhu, "The Role of Law and Reform in Financial Markets: The Case of the Emerging

Chinese Securities Market," in *Law Reform and Financial Markets*, ed. Kern Alexander and Niamh Moloney, 145–194 (Cheltenham, UK: Edward Elgar, 2011).

18. In 1995, the government took more drastic action after a high-profile failure when Shanghai International Securities, the country's biggest brokerage, flooded the futures market with sell orders in a brazen attempt to profit from the fall. Kahn, "China Ends Clean-Up Campaign"; Seth Faison, "Trader Sentenced in China Bond Scandal," *New York Times*, February 4, 1997, http://www.nytimes.com/1997/02/04/business/trader -sentenced-in-china-bond-scandal.html; "China Reopens Government Bond Futures Market," *Financial Times*, September 6, 2013, accessed October 30, 2014, www.ft.com/intl/cms/s/0/94600a40-16cf-11e3-9ec2-00144fea bdc0.html; Helen Sun, "China to Start Bond Futures Trading Next Week after 18-Year Halt," *Bloomberg News*, August 30, 2013, accessed October 30, 2014, www.bloomberg.com/news/2013-08-30/china-to-start-bond-futures -trading-next-week-after-18-year-halt.html; Douglas J. Elliott and Kai Yan, "The Chinese Financial System: An Introduction and Overview," Brookings Institution, July 2013, http://www.brookings.edu/~/media/research/files /papers/2013/07/01-chinese-financial-system-elliott-yan/chinese-financial -system-elliott-yan.pdf; Kahn, "China Ends Clean-Up Campaign."

19. Jiang Yang, "Speech by CSRC Vice Chairman Jiang Yang," Caixin Summit China Securities Regulatory Commission, December 19, 2013, www .csrc.gov.cn/pub/csrc_en/Informations/phgall/201402/t20140211_243679 .html.

20. Anonymous trader, interview by David Abraham, February 14, 2014.

21. Agnieszka Troszkiewicz and Maria Kolesnikova, "HKEx Shares Tumble as LME Purchase May Clear Regulator," *Bloomberg News*, June 18, 2012, accessed October 30, 2014, www.bloomberg.com/news/2012-06-17 /hong-kong-lme-bid-faces-regulator-as-32-billion-in-deals-killed.html.

22. HKEx, "HKEx Makes Recommended Cash Offer for the London Metal Exchange," press release, June 15, 2012, accessed October 30, 2014, www.hkex.com.hk/eng/newsconsul/hkexnews/2012/Documents/120615 news.pdf. "The LME deal is so far acting as a drag on profits for HKEx, as the London exchange's contribution of a HK$326 million profit in 2013 was offset by its addition of operating expenses of HK$783 million." Lawrence White and Saikat Chatterjee, "Rpt-Update 2: Hkex Bets on Tech Spend, Yuan Push As 2013 Profits Disappoint," Reuters, February 26, 2014, http://www.reuters.com/article/2014/02/26/hkex-earnings-idUSL3N 0LQ13E20140226.

23. Organisation de Coopération et de Développement Économiques, "Steelmaking Raw Materials: Market and Policy Developments," October 11, 2012, accessed October 30, 2014, http://www.oecd.org/sti/ind/steelmaking -raw-materials.pdf; Tao Xu, *The Development and Restructuring of China's Metal Mining Industry* (Kingston, ON: Centre for Resource Studies, Queen's University, 1991). China publishes its export quotas and taxes, but some traders complain about minimum prices they must pay at the border.

24. Zhang Yan and Qian Wang, "Smuggling Blights Rare Earths Industry," *China Daily,* December 10, 2012, accessed October 30, 2014, usa .chinadaily.com.cn/epaper/2012-12/10/content_16002928.htm.

25. Metal-Pages, "China Cuts Indium Export Duty to 2% from 5%," December 17, 2013, accessed October 30, 2014, www.metal-pages.com/news /story/75966/china-cuts-indium-export-duty-to-2-from-5.

26. Peter Foster, "Rare Earths: Why China Is Cutting Exports Crucial to Western Technologies," *Telegraph,* March 19, 2011, accessed October 30, 2014, www.telegraph.co.uk/science/8385189/Rare-earths-why-China -is-cutting-exports-crucial-to-Western-technologies.html.

27. Iin P. Handayani, "Beyond Statistics of Poverty," *Jakarta Post,* February 13, 2012, accessed October 30, 2014, www.thejakartapost.com /news/2012/02/13/beyond-statistics-poverty.html; Joe Cochrane, "New Leader Takes Oath of Office in Indonesia," *New York Times,* October 20, 2014, accessed October 30, 2014, www.nytimes.com/2014/10/21/world/asia /joko-widodo-is-sworn-in-as-indonesian-president.html.

28. Melanie Burton, "China Boosts Tin Ore Exports from Myanmar as Indonesian Supply Dries Up," Shanghai Metals Market, September 9, 2013, accessed October 30, 2014, www.metal.com/newscontent/52859_china -boosts-tin-ore-exports-from-myanmar-as-indonesian-supply-dries-up; Anonymous source, interview by David Abraham, September 8, 2013.

29. Reuters, "Tin Miners Demand Revised Indonesian Mining Law," published on Shanghai Metals Market site, February 22, 2010, accessed October 20, 2014, www.metal.com/newscontent/6775_tin-miners-demand -revised-indonesian-mining-law; Matteo Fagotto, "Indonesia: Tin Men," *Caravan,* June 1, 2014, accessed October 30, 2014, http://www.caravan magazine.in/letters/indonesia-tin-men.

30. Kyodo News International, "Toyota Tsusho to Develop Rare Earth Mineral Business in Indonesia," Free Library, November 19, 2010, accessed October 30, 2014, http://www.thefreelibrary.com/Toyota+Tsusho+to +develop+rare+earth+mineral+business+in+Indonesia.-a0242607002; Linda Yulisman, "Govt to Develop Rare Earth Processing," *Jakarta Post,*

July 11, 2014, accessed October 30, 2014, www.thejakartapost.com/news/2014
/07/11/govt-develop-rare-earth-processing.html.

31. A number of rebel groups are involved in the trade, including
those tied to Rwanda's Hutu militia, which was responsible for the Rwan-
dan genocide in 1994. Cecilia Jamasmie, "Tantalum: A Bloody Future?"
mining.com, September 1, 2009, accessed October 30, 2014, www.mining
.com/tantalum-a-bloody-future-52898.

32. Patrick Stratton and David Henderson, "Tantalum Market Overview," Minor Metals Trade Association, January 31, 2013, accessed October 30, 2014, http://www.mmta.co.uk/tantalum-market-overview.

33. Ibid.; "Tantalum: Raw Materials and Processing," Tantalum-
Niobium International Study Center, accessed October 30, 2014, http://
tanb.org/tantalum.

34. Michael Smith, "How Colombian FARC Terrorists Mining Tungsten Are Linked to Your BMW Sedan," *Bloomberg News*, August 8, 2013,
accessed October 30, 2014, www.bloomberg.com/news/2013-08-08/terrorist
-tungsten-in-colombia-taints-global-phone-to-car-sales.html; Martha
Crenshaw, "Revolutionary Armed Forces of Colombia—People's Army,"
Mapping Militant Organizations, Stanford University, updated July 29,
2014, accessed October 30, 2014, web.stanford.edu/group/mappingmilitants
/cgi-bin/groups/view/89.

35. Smith, "How Colombian FARC Terrorists Mining Tungsten Are
Linked."

36. Philips Corporation, *Annual Report*, 2012, http://www.philips
.com/shared/assets/Investor_relations/pdf/Annual_Report_English_2012
.pdf. As Compaq spokesman, Arch Currid told industry press, "Most of the
components that we get [come] from third-party providers, so where they
get their raw goods is hard to determine." Jamasmie, "Tantalum: A Bloody
Future?"; Lisa Reisman, interview by Brian Sylvester, "The Conflict over
Conflict Metals: Lisa Reisman," *Metals Report*, April 6, 2013, accessed October 30, 2014, www.theaureport.com/pub/na/15333; Apple, Inc., Form SD
2014: Conflict Minerals Report, 2014, http://cdn0.vox-cdn.com/assets
/4531041/AppleConflictMinerals.pdf.

37. Randolph Kirchain, interview by David S. Abraham, Cambridge,
MA, February 22, 2013.

38. Caroline Winter, "Why It's Hard to Make an Ethically Sourced
Smartphone," *Bloomberg Businessweek*, September 20, 2013, accessed
April 12, 2015. http://www.bloomberg.com/bw/articles/2013-09-20/why-its
-hard-to-make-an-ethically-sourced-smartphone; Victoria Knowles, "How

Fairphone is dialed in on supply chain sustainability," *Greenbiz*, August 13, 2014, accessed April 12, 2015, http://www.greenbiz.com/blog/2014/08/13/how-fairphone-dialed-supply-chain-sustainability; Karien Strouken, "Choosing the right materials for your Fairphone," Fairphone Blog, November 5, 2012, http://www.fairphone.com/2012/11/05/choosing-the-right-materials-for-your-fairphone/.

39. European Union/European Commission, "Setting up a Union System for Supply Chain Due Diligence Self-Certification of Responsible Importers of Tin, Tantalum and Tungsten, Their Ores, and Gold Originating in Conflict-Affected and High-Risk Areas," Proposal for a Regulation of the European Parliament and of the Council, Brussels, March 3, 2014, accessed October 30, 2014, http://trade.ec.europa.eu/doclib/docs/2014/march/tradoc_152227.pdf.

40. Enough Project, "2012 Conflict Minerals Company Rankings," Raise Hope for Congo, 2012, accessed October 30, 2014, www.raisehopeforcongo.org/companyrankings.

41. Lisa Reisman, e-mail correspondence, December 30, 2013; Patrick Stratton, "Outlook for the Global Tantalum Market," paper presented at the Metal Events Ltd. Second International Tin and Tantalum Seminar, New York, December 11, 2013.

42. Cecilia Jamasmie, "Illegal Mining Is Latin America's New Cocaine," mining.com, December 23, 2013, accessed October 30, 2014, www.mining.com/illegal-mining-is-latin-americas-new-cocaine-73332/.

43. Yossi Sheffi, interview by David Abraham, Cambridge, MA, February 22, 2012.

44. Boeing, "Boeing and Russian Technologies/VSMPO-AVISMA Sign Titanium Agreement," Boeing News Releases/Statements, June 24, 2010, accessed October 30, 2014, boeing.mediaroom.com/index.php?s=20295&item=1278; Lisa Reisman, "HC Starck Changes Game with Conflict-Free Tungsten Supply," *MetalMiner*, August 6, 2013, accessed October 30, 2014, agmetalminer.com/2013/08/06/hc-starck-changes-game-with-conflict-free-tungsten-supply. Even further upstream, Toshiba signed a deal with Kazakhstan's Kazatomprom in 2011, to establish a joint venture to produce rare metals including rare earths and rhenium. "Toshiba and Kazatomprom Establish JV for Rare Metals," Toshiba Corporation Press Release, September 29, 2011, accessed October 30, 2014, www.toshiba.co.jp/about/press/2011_09/pr2901.htm; Molycorp, "Molycorp, Daido Steel, & Mitsubishi Corporation Announce Joint Venture to Manufacture Sintered NdFeB Rare Earth Magnets," Molycorp News Release, November 28, 2011, accessed October 30, 2014, www.molycorp.com/molycorp-daido-steel-mitsubishi

-corporation-announce-joint-venture-to-manufacture-sintered-ndfeb-rare
-earth-magnets.

45. Statement of Robert Jaffe, professor of theoretical physics, Massachusetts Institute of Technology, *Strategic and Critical Minerals Policy Domestic Minerals Supplies and Demands in a Time of Foreign Supply Disruptions: Oversight Hearing before the Subcommittee on Energy and Mineral Resources of the Committee on Natural Resources*, U.S. House of Representatives, 112th Congress, First Session, Tuesday, May 24, 2011 (Washington, DC: General Printing Office, 2011), available at www.gpo.gov/fdsys/pkg/CHRG-112hhrg66649/html/CHRG-112hhrg66649.htm.

46. *Energy Critical Elements: Identifying Research Needs and Strategic Priorities: Hearing Before the Subcommittee on Energy and Environment Committee on Science, Space, and Technology*, U.S. House of Representatives 112th Congress, First Session, on S. 112-56, December 7, 2011 (Washington: General Printing Office, 2011), accessed October 30, 2014, available at www.gpo.gov/fdsys/pkg/CHRG-112hhrg72167/html/CHRG-112hhrg72167.htm.

6

Tech Needs

1. Egyptians were known to use toothpaste as early as 5000 B.C.; Colgate Oral and Dental Health Center, "History of Toothbrushes and Toothpastes," June 12, 2006, www.colgate.com/app/CP/US/EN/OC/Information/Articles/Oral-and-Dental-Health-Basics/Oral-Hygiene/Brushing-and-Flossing/article/History-of-Toothbrushes-and-Toothpastes.cvsp; Valerie Strauss, "Ever Wondered How People Cleaned Their Teeth before They Had Toothbrushes?" *Washington Post*, April 12, 2009, www.washingtonpost.com/wp-dyn/content/article/2009/04/12/AR2009041202655.html; Colgate, "Brushing & Flossing: Use, Replacement & Technique of the Right Dental Products | Colgate® Oral Care Articles," 2014, http://www.colgate.com/app/CP/US/EN/OC/Information/Articles/Oral-and-Dental-Health-Basics/Oral-Hygiene/Brushing-and-Flossing/article/History-of-Toothbrushes-and-Toothpastes.cvsp; American Dental Association, "Mouth Happy: Toothbrushes," www.ada.org/en/Home-MouthHealthy/az-topics/t/toothbrushes; Philips, "Building a Global Oral Healthcare Brand," Annual Report 2011, www.annualreport2011.philips.com/content_ar-2011/en/proofpoints/global_oral_healthcare_brand.aspx; Eben Pingree, Haley Gilbert, Alex Nadas, Ciatlin Blodget, and Dan Esdorn, "The Electric Toothbrush:

Ecosystem Management Lessons in Consumer Products," faculty.tuck
.dartmouth.edu/images/uploads/faculty/ron-adner/21The_Electric
_Toothbrush.pdf. The price of the toothbrush dropped with the advent of
new product offerings. A basic spinning version was invented in 1939, but
never sold well commercially. Broxo, "The Leaders in Oral Health since
1956," www.broxo.com/about-us.
   2. Ralf Hoppe, "Globalization: The Global Toothbrush," *Der Spiegel,*
January 31, 2006, www.spiegel.de/international/spiegel/globalization-the
-global-toothbrush-a-398229.html.
   3. The Japanese toothbrush did not use cadmium as Sonicare did.
Japanese have an aversion to using cadmium due to pollution, which led to
health effects in the 1930s. Hoppe, "Globalization"; Philips, "Why Sonicare:
Results You Can See and Your Patients Can Feel," www.sonicare.com
/professional/en_us/WhySonicare/WhySonicTechnology.aspx; "Sonicare
Features: Types of Batteries / Multi-voltage Chargers / Traveling Cases,"
www.animated-teeth.com/electric_toothbrushes/b-sonicare-battery
-charger.htm; U.S. Geological Survey and the U.S. Department of the Interior, "U.S. Mineral Commodity Summaries 2012," minerals.usgs.gov
/minerals/pubs/mcs/ and minerals.usgs.gov/minerals/pubs/mcs/2012/mcs
2012.pdf. Japanese government research estimates that thirty-five metals
are needed in an electric toothbrush, about twenty-four of them are minor
metals. Shinsuke Murakami, e-mail, November 25, 2014; Ryan Castilloux,
interview by Gareth Hatch, "Rare Earth Market Outlook: Supply, Demand
and Pricing From 2014-2020," November 28, 2014, http://www.techmet
alsresearch.com/rare-earth-market-outlook-report.
   4. Hayden Dingman, "The Legend of Jack Kilby: 55 Years of the Integrated Circuit," *PCWorld,* September 12, 2013, www.pcworld.com/article
/2048664/the-legend-of-jack-kilby-55-years-of-the-integrated-circuit
.html; Texas Instruments, "What If He Had Gone on Vacation," www.ti
.com/corp/docs/kilbyctr/vacation.shtml.
   5. Nobelprize.org, "The History of the Integrated Circuit," May 5,
2003, www.nobelprize.org/educational/physics/integrated_Circuit/history/;
Dingman, "The Legend of Jack Kilby."
   6. Jack took a sliver of germanium to act as a "semi-conducting" material and affixed it with what looks like a thin layer of used chewing gum
to a small rectangular microscope glass with wires emanating from it. No-
belprize.org, "The History of the Integrated Circuit"; Nobelprize.org, Nobel
Media AB 2014, May 5, 2003, www.nobelprize.org/educational/physics/integ
rated_Circuit/history/, accessed April 6, 2015; Jack S. Kilby–Biographical,
Nobelprize.org, Nobel Media AB 2014, in *Les Prix Nobel: The Nobel Prizes*

2000, ed. Tore Frängsmyr (Stockholm: Nobel Foundation, 2001), available at www.nobelprize.org/nobel_prizes/physics/laureates/2000/kilby-bio.html; Dingman, "The Legend of Jack Kilby."

7. David Rowlands, "The Push-Button Abacus," *Design* 290 (February 1973): 36–37, available at vads.ac.uk/diad/article.php?title=2&year=37 &article=d.290.27.

8. Alice Rawsthorn, "Farewell, Pocket Calculator?" *New York Times*, March 4, 2012, www.nytimes.com/2012/03/05/arts/design/farewell-pocket -calculator.html.

9. Viktor Mayer-Schönberger and Kenneth Cukier, *Big Data: A Revolution that Will Transform How We Live, Work, and Think* (Boston: Houghton Mifflin Harcourt, 2013).

10. Rowlands, "The Push-Button Abacus"; David A. Bell, "Napoleon in the Flesh," *MLN* 120, no. 4 (September 2005): 711–715; "Pocket Calculators Add up to a Big Market," *New Scientist*, Google Books, July 20, 1972, books.google.co.uk/books?id=iRLS1Pz9xJwC&pg=PA144#v=onepage&q &f=false; Walter Alcorn, telephone interview by David Abraham, January 30, 2014.

11. Worlds of Wonder, Teddy Ruxpin online, www.teddyruxpinonline .com/worldsofwonder.html; Douglas Martin, "Ken Forsse, Who Brought a Toy Bear to Life, Dies at 77," *New York Times*, April 8, 2014, www.nytimes .com/2014/04/09/business/ken-forsse-who-brought-a-toy-bear-to-life-dies -at-77; Michael Pollick, "This Fuzzy-Wuzzy Takes Messages 'Wabi,'" *Herald Tribune*, September 11, 2003, www.heraldtribune.com/article/20030911 /NEWS/309110400; Walter Alcorn, telephone interview, January 30, 2014.

12. Linda Wells, "The Toys at the Top of Children's Gift Lists," *New York Times* November 29, 1986, http://www.nytimes.com/1986/11/29/style /the-toys-at-the-top-of-children-s-gift-lists.html; TheCHIVE, "The Best Selling Christmas Gifts from 1980 to 2011 (31 Photos)," December 17, 2012, thechive.com/2012/12/17/the-best-selling-christmas-gifts-from-1980-to -2011-31-photos; "The Most Popular Christmas Toys by Year since 1960," Geek in Heels, November 4, 2009, www.geekinheels.com/2009/11/04/the -most-popular-christmas-toys-by-year-since-1960.html; Amber F. Pietrobono, interview by Joanna Stern, New York, February 12, 2013, Joanna Stern, "At Toy Fair 2013, the iPad Is the Real Plaything," ABC News, February 12, 2013, http://abcnews.go.com/Technology/barbie-digital-makeover-mirror -idollhouse-nerf-cyberhoops-toy/story?id=18477909; Stern, "Supertoy Teddy: The Teddy Ruxpin for the Smartphone and Siri Generation," ABC News, July 29, 2013, abcnews.go.com/Technology/supertoy-teddy-teddy -ruxpin-smartphone-siri-generation/story?id=19806856.

13. K. Hodgkins, "IntoMobile Perspective: How Motorola Went from the Car, to the Moon and Finally Your Pocket," IntoMobile, October 2, 2014, www.intomobile.com/2012/10/02/intomobile-perspective-motorola-then -and-now; Motorola, "What Better Way to Discover Motorola's Heritage Than by Exploring the Stories Behind Some of Our Biggest Innovations?" www.motorola.com/us/consumers/about-motorola-us/About_Motorola -History-Timeline/About_Motorola-History-Timeline.html; "History of the Cellular (Cell/Mobile) Phone - Handsets - DynaTAC 8000X," History Cell (Mobile) Phone, www.historyofthecellphone.com/phones/motorola -dynatac-8000x.php; Ellen MacArthur Foundation, "In Depth—Mobile Phones," 2014, http://www.ellenmacarthurfoundation.org/business/toolkit /in-depth-mobile-phones; Whatsinmystuff.org, "Key Facts—What's In My Stuff," 2014, http://www.whatsinmystuff.org/key-facts/; Environmental Protection Agency, "The Life Cycle of a Cell Phone," 2004, www.epa.gov/osw /education/pdfs/life-cell.pdf. A phone is made of roughly 40 percent plastic; the rest is metals and ceramics.

14. Because of its electrical conductivity and spring "memory"—the ability to quickly recoil or snap back to an original formed shape—beryllium serves as the connector between a mobile phone and a charger. "Beryllium: The Miracle Metal," *Materion*, materion.com/~/media/Files/PDFs/Cor porate/Beryllium-The%20Miracle%20Metal; Prachi Patel-Predd, "The Trouble with Touch Screens," *IEEE Spectrum*, January 1, 2009, spectrum .ieee.org/consumer-electronics/gadgets/the-trouble-with-touch-screens; Akio Shibata, "Marubeni Corporation: A View of Rare Metals in the Natural Resource Market," Marubeni Research Institute Presentation, 2007; National Academies Press, "Minerals, Critical Minerals, and the U.S. Economy," 2008, www.nap.edu/catalog/12034/minerals-critical-minerals-and -the-us-economy; Claire Cain Miller, "More Power to You," *Forbes*, July 3, 2008, www.forbes.com/forbes/2008/0721/048b.html.

15. Fuji Film, "Value from Innovation," Annual Report 2014, www .fujifilmholdings.com/en/investors/annual_reports/2013/feature/v80 _keybusinesses.html; Roger Entner, "Roger's Recon: State of Wireless Union 2014, Part Two," *Recon Analytics*, February 13, 2014, reconanalytics.com /2014/02/rogers-recon-state-of-wireless-union-2014-part-two. Oko Institute estimates roughly 6.3 grams of cobalt are in each smartphone battery. Gartner research estimates roughly 1.2 billion smartphones sold in 2014. Matthias Buchert, Andreas Manhart, Daniel Bleher, and Detlef Pingel, "Recycling Critical Raw Materials from Waste Electronic Equipment," e-report, Darmstadt, Oeko-Institut e.V., 2010, http://www.oeko.de/oekodoc /1375/2012-010-en.pdf; Gartner, "Gartner Says Sales of Smartphones Grew

20 Percent in Third Quarter of 2014," 2014, http://www.gartner.com
/newsroom/id/2944819; Whatsinmystuff.org, "Key Facts."

16. Joel Santo Domingo, "SSD vs. HDD: What's the Difference?" *PC Magazine*, February 20, 2014, www.pcmag.com/article2/0,2817,2404258,00
.asp. Molycorp, the company that makes a large percentage of magnets for
hard drives, refers to them as penny magnets due to their low cost. Stan
Trout, telephone interview by David Abraham, February, 5, 2014; Suzanne
Shaw and Steve Constantinides, "Permanent Magnets: The Demand for
Rare Earths," paper presented at the eighth International Rare Earths Conference, Hong Kong, November 13–15, 2012.

17. Steve Constantinides, e-mail, June 13, 2014; Steve Constantinides,
unpublished research, 2014; Santo Domingo, "SSD vs. HDD."

18. Virgin Atlantic, "In Flight Entertainment, Fact Sheet," www
.virgin-atlantic.com/tridion/images/factsheet_ife_tcm4-426058.pdf. It was
just over twenty years ago (1991) that Virgin Atlantic became the first airline
to offer seatback screens throughout the plane. Sam Schechner, "Airlines
Entertain Tablet Ideas," *Wall Street Journal*, September 27, 2012, http://www
.wsj.com/articles/SB10000872396390443916104578020601759253578.

19. Screens are a patented mix of layers of glass, indium, plastics,
and wires that transmit the electric current. Caleb Denison, "OLED vs.
LED | Digital Trends," *Digital Trends*, August 20, 2013, www.digitaltrends
.com/home-theater/oled-vs-led-which-is-the-better-tv-technology/; EIZO
GLOBAL, "How Can a Screen Sense Touch? A Basic Understanding of
Touch Panels," www.eizo.com/global/library/basics/basic_understanding
_of_touch_panel/.

20. Carol Gowans, "Indium Tin Oxide," Indium Corporation Blogs,
February 4, 2013, blogs.indium.com/blog/indium-tin-oxide; John D.
Jorgenson, "Indium," U.S. Geological Survey, 2002, minerals.usgs.gov
/minerals/pubs/commodity/indium/indimyb02.pdf. By 2018, IHs Global,
a research firm, estimates that indium will have a 66 percent share of the
market; see http://press.ihs.com/press-release/design-supply-chain-media
/itos-dominance-touch-screens-challenged-alternative-technolo; Fuji Film,
"Value from Innovation," highly functional materials section. Growth is
from 2012 through 2018.

21. U.S. Census Bureau, "World Population: Total Midyear Population for the World: 1950–2050," www.census.gov/population/international
/data/worldpop/Table_population.php.

22. Fujitsu Laboratories, "Fujitsu Develops Transmitter Power Amplifier Circuit Technology with Industry-Leading Power Efficiency," September 26, 2013, www.fujitsu.com/global/news/pr/archives/month/2013

/20130926-01.html; National Academies Press, "Minerals, Critical Minerals, and the U.S Economy"; National Institute of Standards and Technology, "Overcoming Barriers to Innovation," February 14, 2012, www.nist.gov/public_affairs/factsheet/overcomingbarriers.cfm. Japanese research led to advances in barium titanite oxide applications.

23. Umicore, "High Purity Germanium," eom.umicore.com/en/high-purity/products/highPurityGermanium/. According to Umicore, "Germanium tetrachloride GeCl$_4$ is used as a doping agent in the core of a silica glass optical fiber to increase the refraction index, allowing guidance of optical data through the fiber. To support optimal transmission, a high purity GeCl$_4$ is critical." W. C. Butterman and J. D. Jorgenson, "Mineral Commodity Profiles: Germanium," Open-File Report 2004-1218 (Reston, VA: U.S. Geological Survey, 2005); Raw Materials Supply Group, Annex V to the Report of the Ad-hoc Working Group on Defining Critical Raw Materials, July 30, 2010, https://ec.europa.eu/eip/raw-materials/en/community/document/annex-v-report-ad-hoc-working-group-defining-critical-raw-materials. The EU predicts the market will grow eightfold from 2012 to 2030.

24. Mark Scott, "Emerging Markets Expected to Drive Device Sales," *New York Times,* January 7, 2014, bits.blogs.nytimes.com/2014/01/07/emerging-markets-expected-to-drive-device-salesDisplaySearch, "Global Tablet PC Shipments to Reach 455 Million by 2017," www.displaysearch.com/cps/rde/xchg/displaysearch/hs.xsl/140206_global_tablet_pc_shipments_to_reach_455_million_by_2017.asp. By 2017, it is predicted that 455 million tablets will come off the assembly.

25. Jake Smith, "Flagship Beijing Apple Store Cancels Selling iPhone 4S after Fight Breaks Out," 9 To 5 Mac Apple Intelligence, January 12, 2012, http://9to5mac.com/2012/01/12/flagship-beijing-apple-store-halts-selling-iphone-4s-after-fight-breaks-out/; Scott Baker, "Customers Left Bloody in Scuffle at Apple Store iPad Launch in Beijing," May 8, 2011, http://www.theblaze.com/stories/customers-left-bloody-in-scuffle-at-apple-store-ipad-launch-in-beijing.

26. CNN Wire Staff, "9 on Trial in China over Teenager's Sale of Kidney for iPad and iPhone," August 10, 2012, http://edition.cnn.com/2012/08/10/world/asia/china-kidney-ipad-trial.

27. Indonesia Investments, "Internet Penetration in Indonesia: Rising but Slower than Wanted," January 16, 2014, www.indonesia-investments.com/news/todays-headlines/internet-penetration-in-indonesia-rising-but-slower-than-wanted/item1524; SlideShare, "Indonesia. The Social Media Capital of the World," December 11, 2013, www.slideshare

.net/OnDevice/indonesia-the-social-media-capital-of-the-world#btn. A previous study quotes Semiocast data from July 2012, Indonesia. Zakir Hussain, "4 Too Many: Indonesian Spiritual Guru Splits Up with Half His 8 Wives," *Straits Times,* May 29, 2013, www.stasiareport.com/the-big -story/asia-report/indonesia/story/4-too-many-indonesian-spiritual -guru-splits-half-his-8-wives; Emma Watson, Twitter post, May 29, 2013, https://twitter.com/EmmaWatsonIndo/status/339996137161383936 /photo/1.

28. Indonesia Investments, "Internet Penetration in Indonesia"; Mariel Grazella, "Mobile Data Use Skyrocketing," *Jakarta Post,* January 22, 2014, www.thejakartapost.com/news/2014/01/22/mobile-data-use-sky rocketing.html; Citizen Lab, "IGF 2013: An Overview of Indonesian Inter-net Infrastructure and Governance," University of Toronto, October 25, 2013, citizenlab.org/2013/10/igf-2013-an-overview-of-indonesian-internet -infrastructure-and-governance/; statistic as of 2012; Metal-Pages, "Boom-ing Sales of Wearable Electronics Creating Markets for Minor Metals and Rare Earths," April 11, 2014, www.metal-pages.com/news/story/78823 /booming-sales-of-wearable-electronics-creating-markets-for-minor -metals-and-rare-earths/; International Data Corporation estimates.

29. T. E. Graedel and J. Cao, "Metal Spectra as Indicators of Devel-opment," *Proceedings of the National Academy of Sciences of the United States* 107, no. 49 (2010): 20905–910, doi:10.1073/pnas.1011019107.

30. Air Canada, "Historical Fleet: Boeing 747-400," www.aircanada .com/en/about/fleet/historical/b747-400.html; Metal-Pages, "New Genera-tion Aircraft Increasingly Titanium Dependent," October 26, 2011, blog .metal-pages.com/2011/10/26/new-generation-aircraft-increasingly -titanium-dependent/; Dawn Hickton, "Commercial Aerospace Demand: International Titanium Association," lecture at the Titanium 2012 Confer-ence and Exhibition, Atlanta, October 8, 2012. According to one estimate titanium use may grow in the aerospace business from 21,300 tons in 2011 to 41,200 tons in 2016. Hickton, lecture, quoted in George M. Bedinger, "2011 Minerals Yearbook: Titanium," U.S. Department of the Interior and U.S. Geological Survey, April 2013, minerals.usgs.gov/minerals/pubs /commodity/titanium/myb1-2011-titan.pdf; Rick Shulte, "Innovation, In-cluding Use of Titanium, Drives Airbus Success," *Titanium Today,* Aero-space ed. (2013): 70–71; Margaret Hunt, "Material Threats to the Titanium Industry," *Titanium Today,* Aerospace ed. (2013): 16–17.

31. National Academies Press, "Minerals, Critical Minerals, and the U.S Economy," 55; Scott Miller, "GE Aviation 'Raising a Third Flag' in S.C.," *GSA Business,* April 12, 2010. www.gsabusiness.com/news/33861-ge

-aviation-lsquo-raising-a-third-flag-rsquo-in-s-c; Désirée E. Polyak, "2011 Minerals Yearbook: Rhenium," U.S. Department of the Interior and U.S. Geological Survey, February 2013, minerals.usgs.gov/minerals/pubs /commodity/rhenium/myb1-2011-rheni.pdf.

32. GE Citizenship, "Rhenium Reduction Program: Using Less of a Rare Mineral," citizenship.geblogs.com/rhenium-reduction-program-using -less-of-a-rare-mineral/; National Academies Press, "Minerals, Critical Minerals, and the U.S Economy," 48.

33. GE Citizenship, "Rhenium Reduction Program." Likewise, LCD screen manufacturers used to apply two layers of indium, but switched to single layers when they found that they received the same performance and the same response. Noah Lehrman, interview by David Abraham, New York, January 8, 2014; Claire Miko, Indium Corporation, telephone interview by David Abraham, February 13, 2014.

34. Steven Duclos, "Energy-Critical Materials," paper presented at Materials Research Society Fall Meeting and Exhibit, Boston, November 25, 2012.

35. Karl Gscneidner, telephone interview by David Abraham, November 12, 2013.

36. Duclos, "Energy-Critical Materials."

37. Boeing Facility Tour, Everett, Washington, July 19, 2013.

38. Boeing, "Boeing Celebrates Delivery of 50Th 747-8," May 29, 2013, http://boeing.mediaroom.com/2013-05-29-Boeing-Celebrates-Delivery-of -50th-747-8; Steven J. Duclos, "Testimony before the Subcommittee on Investigations and Oversight of the House Committee on Science and Technology," February 10, 2010, accessed December 15, 2011, gop.science.house .gov/Media/hearings/oversight10/mar16/Duclos.pdf.

39. Randy Kirchain, interview by David Abraham, Cambridge, MA, February 22, 2013.

40. PricewaterhouseCoopers, "Minerals and Metals Scarcity in Manufacturing: The Ticking Time Bomb," 2011, www.pwc.com/resourcescar city; J. Bonasia, "CONSUMER ELECTRONICS PlayStation 2 Shortage Reflects Changing Industry," *Investor's Business Daily*, December 28, 2000, news.investors.com/technology/122800-348096-consumer-electronics -playstation-2-shortage-reflects-changing-industry.htm; Doug Bartholomew, "Boeing Scrubs New Jets Takeoff," eWeek.com, November 8, 2007, www .eweek.com/c/a/Enterprise-Applications/Boeing-Scrubs-New-Jets -Takeoff/; Walter Pohl, *Economic Geology: Principles and Practice: Metals, Minerals, Coal and Hydrocarbons—Introduction to Formation and Sustainable Exploitation of Mineral Deposits* (Chichester, West Sussex:

Wiley-Blackwell, 2011); John Lasker, "The PlayStation War as the West's Insatiable Appetite for Personal Electronics Continues, So Do Africa's Resource Wars," *Columbia Free Press*, December 19, 2013, columbusfreepress.com/article/playstation-war-west%E2%80%99s-insatiable-appetite-personal-electronics-continues-so-do-africa%E2%80%99s. Coltan jumped nearly tenfold from $40 a pound to $380 and Sony was rumored to have trouble getting its PlayStation to the market because it could not get its hands on the material. Blaine Harden, "A Black Mud from Africa Helps Power the New Economy," *New York Times*, August 12, 2001. www.nytimes.com/2001/08/12/magazine/12COLTAN.html; Helen Vesperini, "Congo's Coltan Rush," BBC News, August 1, 2001, news.bbc.co.uk/2/hi/africa/1468772.stm.

41. John Smith, interview by David Abraham, Fairfield, CT, January 8, 2013.

42. Kirchain, interview, February 22, 2013.

43. Kevin Moore, telephone interview by David Abraham, May 6, 2014.

44. Duclos, "Energy-Critical Materials."

7

Environmental Needs

1. Isa Soares, "Estonia's Dirty Energy Drive for Self-Sufficiency," CNN, December 5, 2013, edition.cnn.com/2013/12/05/business/estonia-oil-shale-industry/.

2. International Energy Agency, "Estonia Is Cleansing Oil Shale," January 2014, www.iea.org/newsroomandevents/news/2014/january/estoniaiscleansingoilshale.html; Linas Jegelevicius, "Baltics' Estonia Ramps Up Wind Power Generation," *Renewable Energy World*, February 2014, http://www.renewableenergyworld.com/rea/news/article/2014/02/baltics-estonia-ramps-up-wind-power-generation. "Elering: Electricity Consumption and Production in Estonia," accessed December 4, 2014, elering.ee/electricity-consumption-and-production-in-estonia-2/.

3. American Physical Society, "Energy Critical Elements," accessed December 4, 2014, www.aps.org/policy/reports/popa-reports/upload/elementsreport.pdf.

4. U.S. Department of Energy, *Critical Materials Strategy*, 2011, energy.gov/sites/prod/files/DOE_CMS2011_FINAL_Full.pdf; Core Writing Team, "Climate Change Synthesis Report Summary For Policymakers,"

Intergovernmental Panel on Climate Change, Fifth Assessment Synthesis Report, 2014, http://www.ipcc.ch/pdf/assessment-report/ar5/syr/SYR_AR5 _SPMcorr1.pdf.

5. If the world wants to maintain carbon emissions below 450 ppm then, according to the International Energy Agency, the world needs more than 15,000 TWh, to be produced from alternative sources. Organization for Economic Cooperation and Development/International Energy Agency, "Renewable Energy Outlook, Basking in the Sun?" *World Energy Outlook, 2013,* 197–232, available at http://www.worldenergyoutlook.org/media /weowebsite/2013/WEO2013_Ch06_Renewables.pdf; International Energy Agency, *Tracking Clean Energy Progress 2013* (2014), www.iea.org/publica tions/TCEP_web.pdf. We will also need ten million hybrids produced annually. *BP Statistical Review of Global Energy,* June 18, 2014, accessed April 15, 2015, http://www.bp.com/en/global/corporate/about-bp/energy -economics/statistical-review-of-world-energy.html.

6. Elisa Alonso, Andrew M. Sherman, Timothy J. Wallington, Mark P. Everson, Frank R. Field, Richard Roth, and Randolph E. Kirchain, "Evaluating Rare Earth Element Availability: A Case with Revolutionary Demand from Clean Technologies," *Environmental Science and Technology* 46, no. 6 (2012): 3406–14, doi:10.1021/es203518d; Elisa Alonso, interview by David Abraham, Baltimore, May 1, 2014; David Abraham, "Geopolitics and Minor Metals," paper presented at the Minor Metals Metal-Pages Conference, Shanghai, China, September 6, 2013.

7. For reference, 1 gigawatt hour of electricity or 1,000 megawatt hours is the annual electricity requirement of 800,000 homes in the United States. Donald Bleiwas, *Byproduct Mineral Commodities Used for the Production of Photovoltaic Cells* (U.S. Department of the Interior and U.S. Geological Survey, 2011); "Magnet Price Hikes Hit Direct Drive's Future," *Windpower Monthly,* October 28, 2011, http://www.windpowermonthly.com/article /1101036/magnet-price-hikes-hit-direct-drives-future; Bo-Ping Hu, "China's Rare-Earth Permanent Magnet Industry," paper presented at the International Conference on Rare Earths (2010), as referenced in U.S. Department of Energy, *Critical Materials Strategy.* As Alstom notes in December 2013 when putting in the largest wind farm offshore, "Thanks to a permanent-magnet generator, there are less mechanical parts inside the device, making it more reliable and thus helping to reduce operating and maintenance costs." Alstom, "Alstom Installing World's Largest Offshore Wind Turbine off the Belgian Coast," November 2013, http://www.alstom.com/press-centre /2013/11/alstom-installing-worlds-largest-offshore-wind-turbine; Margaret Schleifer, "The Giant Wind Gamble," Cleantech Investor, 2014, http://

www.cleantechinvestor.com/portal/wind-energy/9552-the-giant-wind
-gamble.html.

8. The gearboxes convert the 15–20 rpms into higher speeds of 1,800 rpms, which generates electricity. April Nowicki and Peter Bronski, "Are Direct-Drive Turbines the Future of Wind Energy?" Earthtechling, February 2013, http://earthtechling.com/2013/02/are-direct-drive-turbines-the -future-of-wind-energy/; Frank J. Bartos, "Direct-Drive Wind Turbines Flex Muscles," Control Engineering, 2014, http://www.controleng.com/single -article/direct-drive-wind-turbines-flex-muscles/4be132ffb053a53fc4ab10b 2c9c57340.html; Morris Lindsay, "Direct Drive vs. Gearbox: Progress on Both Fronts," Power Engineering 115, no. 3 (2011), http://www.power-eng .com/articles/print/volume-115/issue-3/features/direct-drive-vs-gearbox -progress-on-both-fronts.html; Shawn Sheng, Jon Keller, and Mark Mc-Dade, "Gearbox Reliability Collaborative Update: A Brief," paper presented at the AWEA Operations and Maintenance Working Group Meeting, January 10–12, 2012, www.nrel.gov/docs/fy12osti/53804.pdf.

9. Estimates are approximations because much information on material is proprietary. Steve Constantinides, e-mail, December 1, 2014; Gareth Hatch, telephone interview by David Abraham, August 12, 2014. Dysprosium is 3–8 percent of the magnet's weight. U.S. Department of Energy, Critical Materials Strategy; Laura DiMugno, "North American Windpower: Smaller And Lighter 10 MW Wind Turbines May Be on the Horizon," North American Windpower, 2013, http://www.nawindpower .com/e107_plugins/content/content.php?content.10990; Sander Hoenderdaal, Luis Tercero Espinoza, Frank Marscheider-Weidemann, and Wina Graus, "Can a Dysprosium Shortage Threaten Green Energy Technologies?" Energy 49 (2013): 344–55.

10. U.S. Department of Energy, Critical Materials Strategy.

11. PricewaterhouseCoopers, "Minerals and Metals Scarcity in Manufacturing: The Ticking Time Bomb," accessed March 3, 2014, www.pwc.com /gx/en/sustainability/research-insights/metal-minerals-scarcity.jhtml/.

12. Hatch, interview, August 12, 2014.

13. "Magnet Price Hikes," Windpower Monthly; Metal-Pages, "Siemens Upgraded Wind Turbine Reflects Industry Concerns over Dysprosium Supplies," 2014, www.metal-pages.com/news/story/77941/siemens -upgraded-wind-turbine-reflects-industry-concerns-over-dysprosium -supplies/; Eize de Vries, "GE Energy to Acquire Offshore Wind Turbine Supplier ScanWind of Norway," August 24, 2009, http://www.renewable energyworld.com/rea/news/article/2009/08/ge-energy-to-acquire-off shore-wind-turbine-supplier-scanwind-of-norway.

14. U.S. Department of Energy, *Critical Materials Strategy*; Gerald Fox, "Gearbox Designers Turn to Hybrid Solutions," *Renewable Energy World,* April 2012, www.renewableenergyworld.com/rea/news/article/2012/04/gearbox-designers-turn-to-hybrid-solutions. Drive train specialist Winergy developed a new geared drive with both permanent magnet synchronous generators and an alternative option with "classic" electrically excited synchronous generator. *2013 Global Gearboxes for Wind Energy Product Differentiation Excellence Award,* 1st ed., e-book (Frost and Sullivan, 2013), available at www.winergy-group.com/root/img/pool/downloads/en/winergy_-_award_write-up_final.pdf.

15. Michael Silver, telephone interview by David Abraham, March 13, 2014.

16. ARPA-E, "Cerium-Based Magnets," 2012, arpa-e.energy.gov/?q=slick-sheet-project/cerium-based-magnets. William McCallum, telephone interview by David Abraham, May 8, 2014; Jim Witkin, "A Push to Make Motors with Fewer Rare Earths," *New York Times,* April 22, 2012, www.nytimes.com/2012/04/22/automobiles/a-push-to-make-motors-with-fewer-rare-earths.html; Catherine Yang, "Rare-Earth Trade Dispute with China Heats Up, Scientists Seek Alternatives," *National Geographic,* March 2012, news.nationalgeographic.com/news/energy/2012/03/120330-china-rare-earth-minerals-energy.

17. Suzanne Shaw and Steve Constantinides, "Permanent Magnets: The Demand for Rare Earths," paper presented at the eighth International Rare Earths Conference, Hong Kong, November 13–15, 2012, http://www.roskill.com/reports/minor-and-light-metals/news/download-roskills-paper-on-permanent-magnets-the-demand-for-rare-earths; Reuters, "U.S. to Require New Cars to Have Rearview Cameras by 2018," March 31, 2014, www.reuters.com/article/2014/03/31/us-usa-autos-visibility-idUSBRE A2U16R20140331. Electric vehicles use $1,000 worth of semiconductors in their drive trains, which help bring the engine's power to the wheels, ten times the amount found in standard cars. Harald Bauer, Jan Veira, and Florian Weig, "Moore's Law: Repeal or Renewal," McKinsey on Semiconductors, no. 13 (Autumn 2013), http://www.mckinsey.com/client_service/semiconductors/latest_thinking/autumn_2013_issue.

18. Steve Constantinides, telephone interview by David Abraham, December 1, 2014. Because the exact amount of material inside an automobile is a closely guarded secret, these figures are estimates; Miller quoted in Witkin, "A Push to Make Motors."

19. David Reeck, telephone interview by David Abraham, April 4, 2014.

20. Ed Becker, telephone interview by David Abraham, April 25, 2014.

21. "PODCAST 2B Malcolm Burwell Interviews Pete Savagian about Induction and Permanent Magnet Motors," video, 2013, youtu.be/XV9kJ VrhqNg.

22. Ibid.; Kevin Moore, telephone interview by David Abraham, March 3, 2014.

23. Becker, interview, April 25, 2014.

24. David C. Mowery and Nathan Rosenberg, *Paths of Innovation* (Cambridge: Cambridge University Press, 1998). At the turn of the century there were 1,681 steam-, 1,575 electric-, and 936 gas-powered vehicles. James J. Flink, *America Adopts the Automobile, 1895–1910* (Cambridge, MA: MIT Press, 1970); Curtis D. Anderson and Judy Anderson, *Electric and Hybrid Cars* (Jefferson, NC: McFarland, 2010); Willis Frederick Dunbar and George S. May, *Michigan: A History of the Wolverine State* (Grand Rapids, MI: Eerdman, 1980).

25. Metal-Pages, "Market for Hybrid & Electric Vehicles to Soar within 10 Years: Opportunity for Metals and Rare Earths," 2013, www.metal -pages.com/news/story/75778/market-for-hybrid-electric-vehicles-to-soar -within-10-years-opportunity-for-metals-and-rare-earths/.

26. Gregory White, "A Mismanaged Palladium Stockpile Was Catalyst for Ford's Write-Off," *Wall Street Journal*, February 6, 2002, wsj.com /news/articles/SB1012944717336886240. The price jumped from $363 in late 1999 to more than $1,082 in 2001. BASF Catalysts, "How Catalytic Converters Work," www.catalysts.basf.com/p02/USWeb-Internet/catalysts/en/content /microsites/catalysts/prods-inds/mobile-emissions/how-it-works.

27. David Hanson, "Critical Materials Problem Continues," *Chemical and Engineering News* 89 (2011), cen.acs.org/articles/89/i43/Critical -Materials-Problem-Continues.html. The estimation is by Ford's Christine Lambert, technical leader.

28. Hanson, "Critical Materials Problem Continues"; White, "A Mismanaged Palladium Stockpile."

29. In the first few years of the 2000s the price of palladium jumped 90 percent; see http://minerals.usgs.gov/minerals/pubs/commodity/plati num/550400.pdf; Kitco, "London Fix Historical Gold," accessed November 4, 2014, http://www.kitco.com/scripts/hist_charts/yearly_graphs.plx. Calculations are based on a loss of $1 billion. Henry E. Hilliard, *Platinum-Group Metals*, U.S. Geological Survey, 2001, minerals.usgs.gov/minerals /pubs/commodity/platinum/550401.pdf. The class action lawsuit accused Ford of not disclosing the buildup in the stockpile of palladium. White, "A Mismanaged Palladium Stockpile."

30. McKinsey, "Battery Technology Charges Ahead," July 2012, http://www.mckinsey.com/insights/energy_resources_materials/battery_technology_charges_ahead; International Energy Agency, *Tracking Clean Energy Progress 2013*; Electrification Coalition, "State of the Plug-in Electric Vehicle Market," 2013, electrificationcoalition.org/sites/default/files/EC_State_of_PEV_Market_Final_1.pdf; Angus MacKenzie, "2013 Motor Trend Car of the Year: Tesla Model S," *Motor Trend Magazine*, January 2013, www.motortrend.com/oftheyear/car/1301_2013_motor_trend_car_of_the_year_tesla_model_s/; Jack Kaskey and Simon Casey, "Tesla to Use North American Material Amid Pollution Worry," Bloomberg, March 28, 2014, www.bloomberg.com/news/2014-03-28/tesla-to-use-north-american-material-amid-pollution-worry.html.

31. Jon Sonneborn, telephone interview by David Abraham, April 4, 2014.

32. Merrill Lynch analyst John Lovallo states that the company believes a range of $100–$150/kWh is needed to create a competitive advantage over internal combustion end vehicles. Chris Ciaccia, "Tesla Unveils Gigafactory: What Wall Street's Saying," Thestreet, February 27, 2014, http://www.thestreet.com/story/12459694/1/tesla-unveils-gigafactory-what-wall-streets-saying.html; Sebastian Anthony, "Tesla's Model S Now Has a Titanium Underbody Shield to Reduce Risk of Battery Fires to 'Virtually Zero,'" *Extremetech*, March 28, 2014, http://www.extremetech.com/extreme/179422-teslas-model-s-now-has-a-titanium-underbody-shield-to-reduce-risk-of-battery-fires-to-virtually-zero.

33. McKinsey,"Battery Technology Charges Ahead"; International Energy Agency, *Tracking Clean Energy Progress 2013*; Kaskey and Casey, "Tesla to Use North American Material"; ARPA-E, "The All-Electron Battery: A Quantum Leap Forward in Energy Storage," 2010, arpa-e.energy.gov/?q=programs/beest.

34. Cyrus Wadia, Paul Albertus, and Venkat Srinivasan, "Resource Constraints on the Battery Energy Storage Potential for Grid and Transportation Applications," *Journal of Power Sources* 196, no. 3 (2011): 1593–98, doi:10.1016/j.jpowsour.2010.08.056.

35. Bill Canis, "Battery Manufacturing for Hybrid and Electric Vehicles: Policy Issues," *Congressional Research Service*, 2013, www.fas.org/sgp/crs/misc/R41709.pdf; Mike Ramsey, "Ford CEO: Battery Is Third of Electric Car Cost," *Wall Street Journal*, April 17, 2012, wsj.com/news/articles/SB10001424052702304432704577350052534072994. Ford's CEO Alan Mulally "indicated battery packs for the company's Focus electric car cost between $12,000 and $15,000 apiece."

36. Alex Teran, telephone interview by David Abraham, February 10, 2014.

37. Wadia, Albertus, and Srinivasan, "Resource Constraints on the Battery Energy Storage," 1597.

38. Kaskey and Casey, "Tesla to Use North American Material"; Gareth Hatch, "Going Natural: The Solution to Tesla's Graphite Problem," Technology Metals Research, March 25, 2014, www.techmetalsresearch .com/2014/03/going-natural-the-solution-to-teslas-graphite-problem; Teran, telephone interview, February 10, 2014; Johnsoncontrols.com, "Start-Stop Battery" n.d., accessed April 10, 2015, http://www.johnsoncon trols.com/content/us/en/about/our_company/featured_stories/start-stop _battery.html.

39. Kevin Bullis, "Solarcity Will Use Tesla's Batteries to Store Solar Power," MIT Technology Review, December 5, 2013, http://www.techno logyreview.com/news/522226/solarcity-using-tesla-batteries-aims-to -bring-solar-power-to-the-masses/.

40. John Sheilds, International Molybdenum Association, "Molybdenum in Power Generation," July, 2013, www.imoa.info/download_files /sustainability/IMOA_solar_15.pdf, accessed April 6, 2015; Robert Jaffe, "Symposium D: Energy-Critical Materials," Fall Meeting of the Materials Research Society, Boston, November 25–30, 2012, www.mrs.org/f12-cfp-d; Donal Bleiwas, "Byproduct Mineral Commodities Used for the Production of Photovoltaic Cells," 2014, U.S. Department of the Interior and U.S. Geological Survey, pubs.usgs.gov/circ/1365/Circ1365.pdf.

41. Wyatt Metzger, interview by David Abraham, Golden, CO, July 11, 2013.

42. Ibid.; U.S. Energy Information Administration, "Levelized Cost and Levelized Avoided Cost of New Generation Resources in the Annual Energy Outlook 2014," www.eia.gov/forecasts/aeo/electricity_generation.cfm.

43. Michael W. George, Mineral Commodity Summaries: Gold, 1st ed., e-book (U.S. Geological Survey, 2013), available at minerals.usgs.gov /minerals/pubs/commodity/gold/mcs-2013-gold.pdf; American Council for An Energy-Efficient Economy, "Elevators," 2014, http://www.aceee.org /topics/elevators; Masato Sagawa, interview by David Abraham, Kyoto, Japan, July 20, 2013; Energy-Efficient Elevator Machines, 4th ed., e-book (Honolulu: ThyssenKrupp Elevator Americas, 2014), available at www.boma .org/research/Documents/Find%20a%20Resource/Elevator%20Modern ization%20Case%20Study.pdf.

44. Energy, "Lighting Choices to Save You Money," November 5, 2014, energy.gov/energysaver/articles/lighting-choices-save-you-money;

Consumer Energy Center, "Incandescent, LED, Fluorescent, Compact Fluorescent and Halogen Bulbs," 2014, www.consumerenergycenter.org /lighting/bulbs.html.

45. U.S. Department of Energy, *Critical Materials Strategy*.

46. Michael Silver, telephone interview by David Abraham, March 13, 2014; Veolia, "Why Are Rare Earth Metals So Important?" September 2013, http://lamprecycling.veoliaes.com/newsletter/September2013/6.

47. *Energy Independence and Security Act of 2007*, Public Law 110-140 (2007), *U.S. Statutes at Large* 121 (2007): 1492; Osram, "Osram Expands with New LED Assembly Plant In China," 2012, http://www.osram.com /osram_com/press/press-releases/_business_financial_press/2012/wuxi/; Vance Cariaga, "LED Lighting Gets a Warm Reception in China; Prices Drop CREE Durham, North Carolina," *Investor's Business Daily*, April 25, 2013; U.S. Department of Energy, *Critical Materials Strategy*; LED News (LED knowledge, technologies, events), "Chinese LED Lighting Market Situation," December 9, 2013, http://www.lednews.org/chinese-led-lighting -market-situation/. China produced more than 80 percent of compact fluorescent bulbs in 2013. David Solomon, "Work Now Done in Manchester Being Moved to Mexico, China," *Union Leader*, March 20, 2014, www .unionleader.com/article/20140320/NEWS02/140329892; Associated Press, "Six Ohio GE Plants to Close," October 6, 2007, www.ohio.com/business /six-ohio-ge-plants-to-close-1.70725.

48. Tadeu Carneiro, interview by David Abraham, Araxá, Brazil, May 14, 2013.

49. International Energy Agency, *Tracking Clean Energy Progress 2013*.

50. Kevin Bullis, "Automakers Shed the Pounds to Meet Efficiency Standards," *MIT Technology Review*, February 20, 2013, www.technolo gyreview.com/news/510611/automakers-shed-the-pounds-to-meet-fuel -efficiency-standards; Renew Economy, "Research: Solar Panel Efficiency Boosted 30%," May 31, 2012, reneweconomy.com.au/2012/research-solar -panel-efficiency-boosted-30-84356; Hitachimetals.com, "HMA Specialty Steel: Hitachi Metals America," accessed on April 5, 2015, www.hitachi metals.com/product/specialtysteel/solarpower.

51. Marta Vieira, "CBMM Invests $1 Billion to Plant in Araxá," *Estado de Minas*, August 13, 2013, http://www.em.com.br/app/noticia/eco nomia/2013/08/13/internas_economia,434415/cbmm-investe-r-1-bilhao -para-fabrica-em-araxa.shtml.

52. Edward Wong, "Air Pollution Linked to 1.2 Million Deaths in China," *New York Times*, April 1, 2013, www.nytimes.com/2013/04/02/world /asia/air-pollution-linked-to-1-2-million-deaths-in-china.html; BBC News,

"Lung Cancer Cases Soar in Beijing," November 9, 2013, www.bbc.com /news/magazine-24880737.

53. Juan Du, "Opportunities Abound in Clean Technology," *China Daily*, August 14, 2013, usa.chinadaily.com.cn/business/2013-08/13/content _16891896.htm; Liu Yuanyuan, "China Increases Target for Wind Power Capacity to 1,000 GW by 2050," *Renewable Energy World*, January 2012, www .renewableenergyworld.com/rea/news/article/2012/01/china-increases -target-for-wind-power-capacity-to-1000-gw-by-2050 $1.9 trillion. China is focusing on developing its strategic industries, most of them green, so they will contribute 15 percent of the country's gross domestic product growth by 2020, up from 5 percent. See David Abraham and Meredith Ludlow, "China's New Plan For Economic Domination," *The Atlantic*, May 18, 2011, http://www.theatlantic.com/international/archive/2011/05/chinas-new -plan-for-economic-domination/239041/; Jianxiang Yang, "China Plans $1.8 Trillion Wind Power Plan for 2050," *Windpower Monthly*, October 20, 2011, www.windpowermonthly.com/article/1099715/china-plans-18-trillion -wind-power-plan-2050.

54. Global Wind Energy Council, "Global Wind Statistics 2013," May 2, 2014, www.gwec.net/wp-content/uploads/2014/02/GWEC-PRstats-2013 _EN.pdf.

8

War Effort

1. Dennis Jenkins, *Lockheed Secret Projects: Inside the Skunk Works* (St. Paul, MN: MBI, 2001); Central Intelligence Agency (CIA), "From Drawing Board to Factory Floor," February 22, 2012, accessed December 16, 2014, www.cia.gov/library/center-for-the-study-of-intelligence/csi-publi cations/books-and-monographs/a-12/from-the-drawing-board-to-factory -floor.html. In 1954, the CIA contracted Lockheed to build a spy plane; see Dennis Jenkins, *Lockheed Secret Projects*.

2. "Air Force Seeks 'Unobtanium,'" *Albuquerque Tribune*, November 9, 1957; John Adams, *Remaking American Security: Supply Chain Vulnerabilities & National Security Risks across the U.S. Defense Industrial Base* (Washington, DC: Alliance for American Manufacturing, 2013), http://www.americanmanufacturing.org/research/entry/remaking -american-security; Ben R. Rich and Leo Janos, *Skunk Works: A Personal Memoir of My Years at Lockheed* (Boston: Little, Brown, 1994); CIA, "Breaking Through Technological Barriers," August 14, 2012, accessed

December 16, 2014, https://www.cia.gov/library/center-for-the-study-of
-intelligence/csi-publications/books-and-monographs/a-12/breaking
-through-technological-barriers.html#4-rich-and-janos.

3. "Air Force Seeks 'Unobtanium,'" *Albuquerque Tribune.*

4. Adams and Paulette, *Remaking American Security*; John
Browne, *Seven Elements that Have Changed the World* (London: Weidenfeld
and Nicolson, 2013), 151; Chemical Elements, "Titanium (Ti)," accessed December 16, 2014, www.chemicalelements.com/elements/ti.html.

5. Paul M. Tyler, "Growth of Titanium Industry," *Metal Progress* 74,
no. 1 (1958): 172; Gregory W. Pedlow and Donald E. Welzenbach, "The Central Intelligence Agency and Overhead X Reconnaissance: The U-2 and
OXCART Programs, 1954–1974," History Staff Central Intelligence Agency,
Washington, DC, 1992, http://www.foia.cia.gov/sites/default/files/docu
ment_conversions/18/1992-04-01.pdf; Browne, *Seven Elements that Have
Changed the World.*

6. In 1953, the Aircraft Industries of America estimated that in the
1950s titanium alloys would make up 3 percent of the plane's weight, rising
to 30 percent over the long-term. A survey conducted by the Battelle Memorial Institute indicated that up to 50 percent may be used. Paul M. Tyler,
"Present and Future Uses of Titanium," *Metal Progress* 74, no. 3 (1958): 178.
Ninety-three percent of the plane's weight was titanium alloy. Peter W.
Merlin, "Mach 3+ NASA USAF YF-12 Flight Research 1969–1979," Washington DC, NASA Historical Division, 2002, p. 3, available at http://www
.nasa.gov/centers/dryden/pdf/88796main_YF-12.pdf.

7. Cathy Cobb and Harold Goldwhite, *Creations of Fire: Chemistry's
Lively History from Alchemy to the Atomic Age* (New York: Plenum Press,
1995); Rolf E. Hummel, *Understanding Materials Science: History, Proper-
ties, Applications,* 2nd ed. (New York: Springer, 2005), 7. The Bronze Age
began around 3000 B.C. Eric H. Cline, *The Oxford Handbook of the Bronze
Age Aegean (ca. 3000–1000 BC)* (New York: Oxford University Press, 2010);
Richard A. Gabriel and Karen S. Metz, *From Sumer to Rome: The Military
Capabilities of Ancient Armies* (New York: Greenwood Press, 1991), 49;
George Rapp Jr. and Christopher Hill, *Geoarchaeology: The Earth-Science
Approach to Archaeological Interpretation,* 2nd ed. (New Haven, CT: Yale
University Press, 2006).

8. Cobb and Goldwhite, *Creations of Fire;* William Hardy McNeill,
The Pursuit of Power: Technology, Armed Force, and Society since A.D. 1000
(Chicago: University of Chicago Press, 1982); J. D. Muhly, R. Maddin,
T. Stech, and E. Ozgen, "Iron in Anatolia and the Nature of the Hittite
Iron Industry," *Anatolian Studies* 35 (1985): 67, available at www.jstor.org

/discover/10.2307/3642872?uid=3738224&uid=2129&uid=2&uid=70&uid =4&sid=21103299383813. About copper in the earth's crust, see U.S. Congress, Office of Technology Assessment, "Copper: Technology and Competitiveness," 1988, https://www.princeton.edu/~ota/disk2/1988/8808 /880801.PDF. About tin in the earth's crust, see Steve Gagnon, "The Element Tin," It's Elemental, accessed December 16, 2014, http://education.jlab .org/itselemental/ele050.html; Terence Wise and Angus McBride, *Ancient Armies of the Middle East* (Oxford: Osprey, 2003), 40.

9. Richard H. Graham, *SR-71: The Complete Illustrated History of the Blackbird, the World's Highest, Fastest Plane* (Minneapolis, MN: Zenith Press, 2013); Rich and Janos, *Skunk Works;* CIA, "Breaking Through Technological Barriers."

10. David Robarge, "Archangel: CIA's Supersonic A-12 Reconnaissance Aircraft," February 2, 2012, https://www.cia.gov/library/center-for -the-study-of-intelligence/csi-publications/books-and-monographs/a-12 /Archangel-2ndEdition-2Feb12.pdf; CIA, "Breaking Through Technological Barriers"; CIA, "Images and Thumbnails," accessed December 16, 2014, www.cia.gov/library/center-for-the-study-of-intelligence/csi-publications /books-and-monographs/a-12/images-and-thumbnails/.

11. Rich and Janos, *Skunk Works;* Peter W. Merlin, "Design and Development of the Blackbird: Challenges and Lessons Learned," paper presented at the forty-seventh AIAA Aerospace Sciences Meeting Including the New Horizons Forum and Aerospace Exposition, 2009, January 5–8, 2009, Orlando, Florida, http://enu.kz/repository/2009/AIAA-2009-1522.pdf; CIA, "Breaking Through Technological Barriers"; Philip Taubman, *Secret Empire: Eisenhower, the CIA, and the Hidden Story of America's Space Espionage* (New York: Simon and Schuster, 2003).

12. Big Bertha was designed to break concrete fortifications. Sam Kean, *The Disappearing Spoon: And Other True Tales of Madness, Love, and the History of the World from the Periodic Table of the Elements* (New York: Little, Brown, 2010); Marc Romanych and Martin Rupp, *42cm "Big Bertha" and German Siege Artillery of World War I* (London: Osprey, 2014); William S. Hendon, *Letters from France* (Martinsville, NC: Lulu, 2010); Else Ury, *Nesthäkchen and the World War: First English Translation of the German Children's Classic,* trans. Steven Lehrer (Lincoln, NE: iUniverse, 2006).

13. Kean, *The Disappearing Spoon;* WebElements, "Molybdenum: The Essentials." Periodic Table, accessed December 16, 2014, www.webelements .com/molybdenum.

14. Howard Rosenberg, "'Great War': Long Ago but Not So Far Away," *Los Angeles Times,* November 8, 1996, accessed December 16, 2014, articles

.latimes.com/1996-11-08/entertainment/ca-64011_1_great-war; Mike Sharp, Ian Westwell, and John Westwood, *History of World War I* (New York: Marshall Cavendish, 2002).

15. Kean, *The Disappearing Spoon.*

16. Colin G. Fink, "Tungsten, the Key Metal," in *Metallurgical & Chemical Engineering,* ed. Eugene Franz Roeber and Howard Coon Parmelee (New York: Electrochemical, 1917), 272; Ronald H. Limbaugh, *Tungsten in Peace and War, 1918–1946* (Reno: University of Nevada Press, 2010), 15.

17. Mildred Gwin Andrews, *Tungsten, the Story of an Indispensable Metal* (Washington, DC: Tungsten Institute, 1955), available at http://www .tungsten.com/Tungsten%20-%20The%20Story%20of%20an%20Indis-pensable%20Metal.pdf.

18. Robert U. Ayres and Benjamin Warr, *The Economic Growth Engine: How Energy and Work Drive Material Prosperity* (Cheltenham, UK: Edward Elgar, 2009), 120; Digital History, "The Espionage Act of 1917," June 15, 1917, accessed December 18, 2014, www.digitalhistory.uh.edu/disp _textbook.cfm?smtID=3&psid=3904; "Plotted to Ship Tungsten," *New York Times,* November 9, 1917, query.nytimes.com/mem/archive-free/pdf?_r=1 &res=9A03EFD9123FE433A2575AC0A9679D946696D6CF.

19. Limbaugh, *Tungsten in Peace and War,* 22; Frank W. Griffin, Hearing before the Committee on Finance, United States Senate, Sixty-Sixth Congress, First Session on HR 4437, a Bill to Provide Revenue for the Government and to Promote the Production of Tungsten Ores and the Manufacture Thereof, November 10–11, 1919 (Washington, DC: Government Printing Office, 1919).

20. Oliver Ralston, the chief metallurgist at the Bureau of Mines, notes that a number of new processes developed during the war hastened the development of the processing industry. Oliver C. Ralston, Stockpile and accessibility of strategic and critical materials to the United States in time of war, Hearings before the Special Subcommittee on Minerals, Materials, and Fuel Economics of the Committee on Interior and Insular Affairs, United States Senate, Eighty-Third Congress, First Session, Pursuant to S. Res. 143, a resolution to investigate the accessibility and availability of supplies of critical raw materials (Washington, DC: Government Printing Office, 1954); Hans Imgrund and Nicole Kinsman, "Molybdenum: An Extraordinary Metal in High Demand," *Stainless Steel World,* 2007, http://www.imoa.info/download_files /molybdenum/Molybdenum.pdf. In the early 1920s, the Wills Saint Claire car was the first to use molybdenum in the steel used in its production.

21. Désirée E. Polyak, *Minerals Yearbook: Molybdenum* (Washington, DC: U.S. Department of the Interior and the U.S. Geological Survey, 2012),

available at http://minerals.usgs.gov/minerals/pubs/commodity/molyb denum/mybl-2011-molyb.pdf. On molybdenum's numerous end uses, see International Molybdenum Association, "Molybdenum Compounds Uses," accessed December 18, 2014, www.imoa.info/molybdenum_uses/moly_che mistry_uses/moly_compounds/molybdenum_compounds.php; Imgrund and Kinsman, "Molybdenum."

22. Shepard B. Clough, *European Economic History* (New York: McGraw-Hill, 1968), 78; Task Force on American Innovation, "Scientific Research," 2012, http://www.innovationtaskforce.org/docs/TFAI%20-%20 Scientific%20Research%202012.pdf.

23. Dominic Boyle, "What Has Been the Effect of the Recent Economic Turndown on Electronic and Minor Metals? Where Is the Market Going?" Lecture, Metal-Pages Conference, Shanghai, September 8, 2013; Dominic Boyle, telephone interview by David Abraham, November 4, 2014.

24. Cool Cosmos, "History of Infrared Astronomy," accessed December 18, 2014, coolcosmos.ipac.caltech.edu/cosmic_classroom/cosmic_refe rence/irastro_history.html.

25. Peter N. Gabby, "Germanium," U.S. Geological Survey, Mineral Commodity Summaries, January 2006, accessed December 18, 2014, http:// minerals.usgs.gov/minerals/pubs/commodity/germanium/germamcs06 .pdf. A 2012 U.S. Geological Survey report indicated that infrared optics made up 50 percent of the market, whereas in 2006 they were roughly 30 percent; see David E. Guberman, "Germanium," U.S. Geological Survey, Mineral Commodity Summaries, January 2012, accessed December 18, 2014, http://minerals.usgs.gov/minerals/pubs/commodity/germanium/mcs -2012-germa.pdf; Michael W. George, "Germanium," U.S. Geological Survey, Mineral Commodity Summaries, January 2005, accessed December 18, 2014, http://minerals.usgs.gov/minerals/pubs/commodity/germanium /germamcs05.pdf; David E. Guberman, "Germanium," in *2009 Minerals Yearbook* (U.S. Geological Survey, 2009).

26. Michael Eckert and Helmut Schubert, *Crystals, Electrons, Transistors: From Scholar's Study to Industrial Research* (New York: American Institute of Physics, 1990).

27. The older diodes burned out quickly. W. C. Butterman and J. D. Jorgenson, "Mineral Commodity Profiles: Germanium," 2005, http://pubs .usgs.gov/of/2004/1218/2004-1218.pdf; Solomon Gartenhaus et al., "A History of Physics at Purdue: The War Period (1941–1945)," Department of Physics and Astronomy, www.physics.purdue.edu/about_us/history/war _period.shtml; Learn About Electronics, "Semiconductor Materials," accessed December 18, 2014, http://www.learnabout-electronics.org/semi

conductors_01.php. Scientists had little use for germanium before the war. It was used for niche markets. In fact, scientists identified it as a treatment for anemia; see George R. Minot and John J. Sampson, "Germanium Dioxide as a Remedy for Anemia," *Boston Medical and Surgical Journal* (1923): 629–632. About germanium as an impurity of lead, see "Oral History Transcript: Dr. Frederick Seitz," interview by Lillian Hoddeson and Paul Henriksen, Rockefeller University, January 26, 1981, http://www.aip.org/history/ohilist /4877_1.html; Eckert and Schubert, *Crystals, Electrons, Transistors.*

28. Learn About Electronics, "Semiconductor Materials"; Doug De-Maw, "Diodes and How They Are Used," March 1985, accessed December 18, 2014, http://www.arrl.org/files/file/Technology/tis/info/pdf/8503034.pdf.

29. Eckert and Schubert, *Crystals, Electrons, Transistors.* Transistors, diodes, and integrated circuits are made from semiconductor materials, often silicon and germanium being the most common. Butterman and Jorgenson, "Mineral Commodity Profiles: Germanium." Germanium diodes were ten times more resistant to burnout. Only a few hundred pounds of germanium were produced annually during the war, but in the early 1950s, forty tons were produced. Bethany Halford, "Germanium," 2003, C&EN: It's Elemental: The Periodic Table, accessed December 18, 2014, pubs.acs.org /cen/80th/print/germanium.html.

30. Butterman and Jorgenson, "Mineral Commodity Profiles: Germanium," 11; Metal-Pages, "Japan Move May Stimulate China Defence Expenditure, Push up Germanium Prices," January 14, 2008, accessed December 18, 2014, www.metal-pages.com/news/story/76382/japan-move -may-stimulate-china-defence-expenditure-push-up-germanium-prices; Sam Freeman et al., *Trends in World Military Expenditure, 2012* (Solna, Sweden: SIPRI, 2013), available at http://books.sipri.org/files/FS/SIPRIFS1304 .pdf. Indonesian, Vietnamese, and Chinese combined expenditures have more than doubled in ten years and are well into the hundreds of billions of dollars. China increased spending in 2014. Edward Wong, "China Announces 12.2% Increase in Military Budget," *New York Times,* March 5, 2014, http://www.nytimes.com/2014/03/06/world/asia/china-military -budget.html?hp&_r=0.

31. "Made in China," *Dan Rather Reports,* episode 213 (2013), Ed Richardson, interviewee, http://www.axs.tv/ui/inc/show_transcripts.php ?ami=A4517&t=Dan_Rather_Reports&en=213.

32. Militaries are no longer dropping large numbers of bombs from planes because they have used precise, targeting missiles since the 1960s when the United States outfitted intercontinental ballistic missiles with integrated circuits. IEEE Global History Network, "Integrated Circuits and

the Space Program and Missile Defense," accessed December 18, 2014, www
.ieeeghn.org/wiki/index.php/Integrated_Circuits_and_the_Space
_Program_and_Missile_Defense.

33. Robert Latiff, telephone interview by David Abraham, January 22, 2014.

34. Redlen Technologies, "About Redlen," accessed December 18, 2014, http://redlen.ca; Valerie Bailey Grasso, "Rare Earth Elements in National Defense: Background, Oversight Issues, and Options for Congress," Congressional Research Service, 2013, http://www.fas.org/sgp/crs/natsec /R41744.pdf; Noah Shachtman, "Hypersonic Cruise Missile: America's New Global Strike Weapon," *Popular Mechanics,* December 4, 2006, www .popularmechanics.com/technology/military/4203874; David Hambling, "In the Military, Toxic Tungsten Is Everywhere," *Wired,* April 21, 2009, www .wired.com/dangerroom/2009/04/toxic-tungste-1; General Dynamics, "25mm Tungsten APFSDS-T," October 28, 2010, www.gd-ots.com/download /25mm%20Tungsten%20APFSDS-T.pdf.

35. For 2012 U.S. net import reliance, see "Mineral Commodity Summaries 2013," U.S. Geological Survey, January 24, 2013, http://minerals.usgs .gov/minerals/pubs/mcs/2013/mcs2013.pdf; "Mineral Commodity Summaries 1996," U.S. Geological Survey, Minerals Information, January 1, 1996, accessed December 18, 2014, minerals.usgs.gov/minerals/pubs/mcs/1996; author estimates. (The figure counts rare earth elements individually.)

36. John Adams and Scott Paul, "Opinion: American Security Should Be Homemade," *Politico,* May 8, 2013, accessed December 18, 2014, www .politico.com/story/2013/05/american-security-should-be-homemade -91017.html#ixzz2uIp2iBY7.

37. The weights of rare earth materials included in the congressional committee report likely include a substantial weight portion from other metals as well. Ronald H. O'Rourke, "Navy Virginia (SSN-774) Class Attack Submarine Procurement: Background and Issues for Congress," Congressional Research Service, July 31, 2014, http://www.fas.org/sgp/crs /weapons/RL32418.pdf. For information on Virginia class submarine purchases, see, "DDG 51 Arleigh Burke Class Guided Missile Destroyer," Defense Acquisition Management Information Retrieval, December 31, 2012, accessed December 18, 2014, http://www.dod.mil/pubs/foi/logistics_mat erial_readiness/acq_bud_fin/SARs/2012-sars/13-F-0884_SARs_as_of_Dec _2012/Navy/DDG_51_December_2012_SAR.pdf. For information on the DDG 51 Aegis Destroyer Ships as of 2012, including expected production until 2016, see "Next Global Positioning System Receiver Equipment," Committee Reports 113th Congress (2013–2014), House Report 113-102,

June 7, 2013, accessed December 18, 2014, thomas.loc.gov/cgi-bin/cpquery/
?&dbname=cp113&r_n=hr102.113&sel=TOC_1021876&; Charley Keyes,
"With Chinese Monopoly, U.S. Should Create Rare Mineral Reserve," CNN
Security Clearance (blog), September 22, 2011, accessed December 18,
2014, http://security.blogs.cnn.com/2011/09/22/with-chinese-monopoly-u
-s-should-create-rare-mineral-reserve/.

38. Adams and Paulette, *Remaking American Security*; Beryllium Science and Technology Association, "Uses & Applications of Beryllium," accessed January 18, 2015, http://beryllium.eu/about-beryllium-and-beryllium
-alloys/uses-and-applications-of-beryllium/; TriQuint, "Triquint Supports
Northrop Grumman in Multi-Nation F-35 / JSF Program," 2009, http://
www.triquint.com/newsroom/news/2009/triquint-supports-northrop
-grumman-in-multi-nation-f-35-jsf-program; "Catalog Update," *Microwave
Journal*, 2009, http://www.microwavejournal.com/articles/print/8243
-catalog; Colin Whelan and Nick Kolias, "Gan Microwave Amplifiers Come
of Age," *Technology Today*, 2010, No 2, http://www.raytheon.com/news
/technology_Today/archive/2010_2.pdf.

39. U.S. Department of Defense, "About the Department of Defense
(DOD)," accessed December 18, 2014, www.defense.gov/about/.

40. Intel, "Intel CEO Brian Krzanich's Keynote: News from CES
2014," accessed December 18, 2014, http://m.intel.com/us/en/events/ces
-2014/press-release-keynote.html; ThinkProgress, "Intel Announces First
'Conflict-Free' Microprocessor," January 7, 2014, accessed December 18,
2014, http://thinkprogress.org/security/2014/01/07/3126271/intel-announces
-launch-conflict-free-microprocessors. Intel made a commitment in 2011.
2011 Intel Corporate Responsibility Report, *Connecting and Enriching Lives
Through Technology*, http://csrreportbuilder.intel.com/PDFFiles/CSR_2011
_Full-Report.pdf.

41. Brajendra Mishra, "Review of Extraction, Processing, Properties
& Applications of Reactive Metals," Proceedings of a Symposium Sponsored
by the Reactive Metals Committee of the Light Metals Division, The Minerals, Metals & Materials Society, Annual Meeting, San Diego, CA, February 28–March 15, 1999, 200.

42. Materials Research Society, "Materials Science for National Defense," www.mrs.org/resources-defense; John Shiffman, "Exclusive: U.S.
Waived Laws to Keep F-35 on Track with China-made Parts," Reuters,
January 3, 2014, www.reuters.com/article/2014/01/03/us-lockheed-f-id
USBREA020VA20140103.

43. Raytheon material scientist, telephone interview by David Abraham, December 3, 2013; Bernard Harris and Kanai Shah, "Detection and

Identification of Radiological Sources," *Technology Today,* 2012, No. 1, http:// www.raytheon.com/news/technology_Today/archive/2012_i1.pdf; Yossi Sheffi, interview by David Abraham, Cambridge, MA, February 22, 2012.

44. Department of Defense, "2014 Climate Change Adaptation Road Map," 2014, www.acq.osd.mil/ie/download/CCARprint.pdf.

9

Sustainable Use

1. "China's Rare Earth Boom Comes at Grim Cost," *China Daily,* April 23, 2012, usa.chinadaily.com.cn/china/2012-04/23/content_15117359.htm.

2. Up to one-third of all the air pollution in China is linked to the manufacturing of exports and an indeterminate amount of localized pollution. J. Lin, D. Pan, S. J. Davis, Q. Zhang, K. He, C. Wang, D. G. Streets, D. J. Wuebbles, and D. Guan, "China's International Trade And Air Pollution in the United States," *Proceedings of the National Academy of Sciences* 111, no. 5 (2014): 1736–41, doi:10.1073/pnas.1312860111. According to the authors, "We find that in 2006, 36% of anthropogenic sulfur dioxide, 27% of nitrogen oxides, 22% of carbon monoxide, and 17% of black carbon emitted in China were associated with production of goods for export."

3. Peter Foster, "Rare Earths: Why China Is Cutting Exports Crucial to Western Technologies," *Telegraph,* March 11, 2011, www.telegraph.co.uk /science/8385189/Rare-earths-why-China-is-cutting-exports-crucial-to -Western-technologies.html; Georgiana A. Moldoveanu and Vladimiros G. Papangelakis, 2013, "Recovery of Rare Earth Elements Adsorbed on Clay Minerals: II. Leaching with Ammonium Sulfate," *Hydrometallurgy,* January 2013, 131–32, 158–66, doi:10.1016/j.hydromet.2012.10.011.

4. Government of China, "Full Text: Situation and Policies of China's Rare Earth Industry," June 20, 2012, www.gov.cn/english/2012-06/20/content _2165802_3.htm. Other sources put the waste at 1,000 kilograms per kilogram of rare earth. Li Fangfang, "Damage of Rare Earth Extraction to the Environment," *Beijing Review,* August 28, 2012, www.bjreview.com.cn /special/2012-08/28/content_478716.htm.

5. Li, "Damage of Rare Earth Extraction."

6. "Rare-Earth Mining in China Comes at a Heavy Cost for Local Villages," *Guardian,* August 7, 2012, www.theguardian.com/environment /2012/aug/07/china-rare-earth-village-pollution.

7. Simon Perry and Ed Douglas, "In China, the True Cost of Britain's Clean, Green Wind Power Experiment: Pollution on a Disastrous Scale,"

March 19, 2011, www.dailymail.co.uk/home/moslive/article-1350811/In
-China-true-cost-Britains-clean-green-wind-power-experiment-Pollution
-disastrous-scale.html#ixzz2ydB3NHSX; Lan Xinzhen, "Rare Earth Resolution," *Beijing Review,* August 13, 2012, www.bjreview.com.cn/business
/txt/2012-08/13/content_474520.htm.

8. Renee Cho, "Rare Earth Metals: Will We Have Enough? State of the
Planet," September 19, 2012, blogs.ei.columbia.edu/2012/09/19/rare-earth
-metals-will-we-have-enough/. "According to the Ministry of Industry and
Information Technology, processing one metric tonne of rare earths produces about seven tonnes of strong acid." CleanBiz Asia, "Chinese Regulator
Hits Back at WTO Rare-Earth Investigation," 2014, accessed December 6,
2014, www.cleanbiz.asia/news/chinese-regulator-hits-back-wto-rare-earth
-investigation#.U0hI_FcvmS; Li Jiabao and Liu Jie, "Rare Earth Industry
Adjusts to Slow Market," *China Daily,* September 7, 2009, www.chinadaily
.com.cn/bw/2009-09/07/content_8660849.htm; Laura Talens Peiró and Gara
Villalba Méndez, "Material and Energy Requirement for Rare Earth Production," *JOM* 65, no. 10 (2013): 1327–40, doi:10.1007/s11837-013-0719-8.

9. Xinzhen, "Rare Earth Resolution"; Kerry Brown, *China and the EU
in Context: Insights for Business and Investors,* e-book (Palgrave Macmil-
lan, 2014), available at www.palgrave.com/page/detail/china-and-the-eu-in
-context-kerry-brown/?K=9781137352385.

10. T. Norgate and S. Jahanshahi, "Low Grade Ores: Smelt, Leach or
Concentrate?" *Minerals Engineering* 23, no. 2 (2010): 65–73, doi:10.1016/j
.mineng.2009.10.002.

11. Nokia, 2013, http://www.nokia.com/global/about-nokia/people
-and-planet/sustainable-devices/products/products/; Apple, "15-inch Mac-
Book Pro with Retina Display, Environmental Report," https://www.apple
.com/id/environment/reports/docs/15inch_MacBookPro_wRetinaDis
play_PER_Oct2013.pdf.

12. Kohmei Halada, "Sustainable Material Use," National Institute
for Materials Science, 2010, www.iges.or.jp/isap/2010/en/pdf/day2/Halada
.pdf; Kohmei Halada, "New Stage of Resource Issues and Strategic Elements Initiative in Japan," Presentation, 2nd International Workshop on
Rare Metals, Boston, Massachusetts, 2012.

13. EJOLT, "Ecological Rucksacks (Hidden Flows)," accessed December 6, 2014, www.ejolt.org/2013/02/ecological-rucksacks-hidden-flows/;
C. Hageluken and C. Meskers, "Complex Lifecycles of Precious and Special
Metals," in *Linkages of Sustainability,* ed. T. E. Graedel and E. van der Voet,
165–197 (Cambridge, MA: MIT Press, 2009). As we know from chapter 4,

since material losses occur at every stage of minor metal production and component formation, only a fraction of what is mined ever makes it to the market, let alone to the end product.

14. Halada, "Sustainable Material Use."

15. U.S. Environmental Protection Agency, *Technical Document: Acid Mining Drainage Prediction*, 1994, water.epa.gov/polwaste/nps/upload/amd .pdf; Ron Cohen, telephone interview by David Abraham, April 10, 2014.

16. Cohen, interview, April 10, 2014.

17. John Robert McNeill, *Something New Under the Sun* (New York: Norton, 2000); SBMC, October 18, 2004, accessed December 6, 2014, www .sbmc.or.jp/english/20041018/Kitakyushu_City_vol2_vol3.htm.

18. Masanori Kaji, "Role of Experts and Public Participation in Pollution Control: The Case of Itai-itai Disease in Japan," *Ethics in Science and Environmental Politics*, no. 12 (2012): 99–111, available at www.int-res.com /articles/esep2012/12/e012p099.pdf. The disease was so traumatic that the town has a museum dedicated to the affliction.

19. Markus A. Reuter and Antoinette van Schaik, "Transforming the Recovery and Recycling of Nonrenewable Resources," *Linkages of Sustainability* (Cambridge, MA: MIT Press, 2009), 149–162, doi:10.7551/mit press/9780262013581.003.0009. Energy to meet total mining needs could rise to 40 percent of all global energy in 2050. Heather L. MacLean et al., "Mineral Resources: Stocks, Flows, and Prospects," working papers in economics, 2010, www.economics.rpi.edu/workingpapers/rpi1003.pdf.

20. Yoshihiko Wada, "A Radioactive Thorium Pollution Case in Malaysia: Asian Rare Earth Incident Revisited," paper presented at the Rare Earth Symposium University of Queensland, Brisbane, Australia, May 31, 2013, www.csrm.uq.edu.au/docs/Yoshi.pdf.

21. Keith Bradsher, "Mitsubishi Quietly Cleans Up Its Former Refinery," *New York Times,* March 9, 2011, www.nytimes.com/2011/03/09/busi ness/energy-environment/09rareside.html?_r=0; Charles Wallace, "Environment: A Question of Pollution and Power: A Tiny Malaysian Village Accuses a Japanese Conglomerate of Dumping Nuclear Waste and Causing Illness," *Los Angeles Times,* December 8, 1992, articles.latimes.com /1992-12-08/news/wr-1751_1_nuclear-waste; Wada, "A Radioactive Thorium Pollution Case."

22. Jessica Elzea Kogel, *Industrial Minerals & Rocks* (Littleton, CO: Society for Mining, Metallurgy, and Exploration, 2009), 788; Yanis Miezitis and Dean Hoatson, "Rare Earths," *Australian Mines Atlas,* 2013, www .australianminesatlas.gov.au/aimr/commodity/rare_earths.html.

23. Keith Bradsher, "Malaysia Gambles on Processing Rare Earths," *New York Times,* www.nytimes.com/2011/03/09/business/energy -environment/09rare.html.

24. Dorothy Kosich, "Molycorp Slapped with Hazardous Waste Fine," *Mineweb,* April 22, 2014, www.mineweb.com/mineweb/content/en /mineweb-industrial-metals-minerals-old?oid=238181&sn=Detail; CBC, "B.C. Mining Giant Admits Polluting U.S. Waters," September 10, 2012, www.cbc.ca/news/canada/british-columbia/b-c-mining-giant-admits -polluting-u-s-waters-1.1177305.

25. It may sound odd but companies need pure water in many ap-plications because it has a significant influence on recoveries. N. J. Shackle-ton, V. Malysiak, D. De Vaux, and N. Plint, "Effect of Cations, Anions, and Ionic Strength on the Flotation of Penrlandite-Pyroxene Mixtures," in *Water in Mineral Processing,* ed. J. Drelich and Jiann-Yang Hwang, 197–210 (Englewood, CO: Society for Mining, Metallurgy, and Explora-tion, 2010).

26. *Super Expensive Metals,* Periodic Table of Videos, 2013, www .youtube.com/watch?v=Fg2WzCzKpYU&list=FLpt3lK3DAjXlA7PMKfu -GMQ.

27. Steve Constantinides, telephone interview by David Abraham, December 1, 2014.

28. "Japan Mulls Move into 'Urban Mining,'" *Australian,* November 4, 2011, accessed December 6, 2014, www.theaustralian.com.au/business /opinion/japan-mulls-move-into-urban-mining/story-e6frg9if-122618 5115681#.

29. Shinsuke Murakami, interview by David Abraham, Tokyo, Japan, October 2, 2014.

30. Statistics on the Management of Used and End-of-Life Electron-ics," accessed December 6, 2014, www.epa.gov/epawaste/conserve/materials /ecycling/manage.htm.

31. PR Newswire, "The Modern Day Recycling Dilemma: How to Safely Sell or Recycle Unwanted Cell Phones and Tablets," April 9, 2014, www.prnewswire.com/news-releases/the-modern-day-recycling -dilemma-how-to-safely-sell-or-recycle-unwanted-cell-phones-and -tablets-254527401.html; E. Damanhuri, "Post-Consumer Waste Recycling and Optimal Production," *Intech,* May 23, 2012, doi:10.5772/2642; www .intechopen.com/books/editor/post-consumer-waste-recycling-and -optimal-production; M. A. Reuter, C. Hudson, A. van Schaik, K. Heis-kanen, C. Meskers, and C. Hagelüken, "Metal Recycling: Opportunities,

Limits, Infrastructure. A Report of the Working Group on the Global Metal Flows to the International Resource Panel," United Nations Environment Programme, 2013, www.unep.org/resourcepanel/Portals/24102/PDFs /Metal_Recycling_Full_Report.pdf; SourceWire, "More Than £1 Billion Worth of Unused Mobile Phones in UK Homes," August 17, 2012, www .sourcewire.com/news/73565/more-than-billion-worth-of-unused-mobile -phones-in-uk.

32. Barbara Reck, telephone interview by David Abraham, May 8, 2013.

33. Reuter et al., "Metal Recycling: Opportunities, Limits, Infrastructure." The largest lithium mines in Chile have an ore grade of .14 percent whereas lithium-ion batteries have about 3.5 percent lithium.

34. Eurobiz Japan, "Urban Mining," October 2013, eurobiz.jp/2013/10/ urban-mining/; Environmental Leader, "Aluminum Can Recycling Rate Hit 67% in 2012," October 30, 2013, www.environmentalleader.com/2013/10 /30/aluminum-can-recycling-rate-hit-67-in-2012/. In the United States, only two of every three aluminum cans are recycled, despite an established infrastructure to handle them. And the cans are relatively easy to reprocess because they are made almost completely of aluminum.

35. Barbara Reck, telephone interview, May 8, 2013.

36. Action and Resource Center, *Investigating the Role of Design in the Circular Economy*, 1st ed., e-book (Actions and Research Centre2013), available at https://www.thersa.org/discover/publications-and-articles /reports/the-great-recovery-exec-summary/.

37. Kenji Baba, Yuzo Hiroshige, and Takeshi Nemoto, "Rare-Earth Magnet Recycling," *Hitachi Review* 62, no. 8 (2013): 452, www.hitachi .com/rev/pdf/2013/r2013_08_105.pdf.

38. Christian Hagelüken and Christina Meskers, "Technology Challenges to Recover Precious and Special Metals from Complex Products," presentation abstract R'09 World Congress, September 14, 2009, Davos, Switzerland, http://ewasteguide.info/files/Hageluecken_2009_R09.pdf.

39. Christian Hagelüken and Christina E. M. Meskers, "Complex Life Cycles of Precious and Special Metals," in Graedel and van der Voet, *Linkages of Sustainability*, 186.

40. Umicore, http://www.umicore.com/en/bu/precious-metals -refining/.

41. This refers to end-of-life recycling. United Nations Development Programme, "Towards an 'Energy Plus' Approach for the Poor: A Review of Good Practices and Lessons Learned from Asia and the Pacific," 2011web

.undp.org/asia/pdf/EnergyPlus.pdf; U.S. Environmental Protection Agency, "Statistics on the Management of Used and End-of-Life Electronics," accessed December 6, 2014, www.epa.gov/epawaste/conserve/materials/ecycling/manage.htm.

42. J. W. Darcy, H. M. Dhammika Bandara, B. Mishra, B. Blanplain, D. Apelian, and Marion H. Emmert, "Challenges In Recycling End-of-Life Rare Earth Magnets," *JOM* 65, no. 11 (2013): 1381–82, doi:10.1007/s11837-013-0783-0.

43. Reuter et al. "Metal Recycling: Opportunities, Limits, Infrastructure."

44. Hagelüken and Meskers, "Complex Life Cycles of Precious and Special Metals," 186.

45. Darcy et al., "Challenges in Recycling."

46. From printed wire boards in electronics. Reuter et al., "Metal Recycling: Opportunities, Limits, Infrastructure."

47. Umicore, "Sustainability," accessed December 6, 2014, http://www.umicore.com/en/vision/our-vision/sustainability/.

48. ScienceDaily, "Researchers Call for Specialty Metals Recycling," September 24, 2012, www.sciencedaily.com/releases/2012/09/120924175211.htm; B. K. Reck and T. E. Graedel, "Challenges in Metal Recycling," *Science* 337, no. 6095 (2012): 690, doi:10.1126/science.1217501.

49. Hagelüken and Meskers, "Complex Life Cycles of Precious and Special Metals," 186; Roy Gordon, "Criteria for Choosing Transparent Conductors," *MRS Bulletin* 25, no. 8 (2000): 52, http://www-chem.harvard.edu/groups/gordon/papers/Gordon_MRS_Bull.pdf, p. 4. For example, coatings on glass can be toxic. Reuter et al., "Metal Recycling: Opportunities, Limits, Infrastructure."

50. Hagelüken and Meskers, "Complex Life Cycles of Precious and Special Metals," 186; Barbara Reck, telephone interview, May 8, 2013.

51. Reuter et al., "Metal Recycling: Opportunities, Limits, Infrastructure."

52. Jelle H. Rademaker, René Kleijn, and Yongxiang Yang, "Recycling as a Strategy against Rare Earth Element Criticality: A Systemic Evaluation of the Potential Yield of Ndfeb Magnet Recycling," *Environmental Science and Technology* 47, no. 18 (2013): 10129–10136, pubs.acs.org/doi/abs/10.1021/es305007w; see especially http://pubs.acs.org/doi/suppl/10.1021/es305007w/suppl_file/es305007w_si_001.pdf, pp. S14–15.

10

The War over the Periodic Table

1. Dominique Patton and Niu Shuping, "UPDATE 1-China Dec Rubber Imports Hit Record Peak on Higher Thai Shipments," Reuters, January 10, 2014, www.reuters.com/article/2014/01/10/china-rubber-imports -idUSL3N0KK0WF20140110.

2. Valerie Bailey Grasso, *Rare Earth Elements In National Defense: Background, Oversight Issues, and Options for Congress* (Congressional Research Service, 2013), fas.org/sgp/crs/natsec/R41744.pdf.

3. David Lague, "China Corners Market in a High-Tech Necessity," *New York Times,* January 22, 2006, www.nytimes.com/2006/01/22 /business/worldbusiness/22iht-rare.html.

4. Chinese Academy of Land and Resource Economics, *A Guide to Investment In China's Mineral Industry (2012),* www.chinaminingtj.org/esp /document/A_Guide_to_Investment_in_China's_Mineral_Industry (2012).pdf; David Barboza, "China Sentences Rio Tinto Employees in Bribe Case," *New York Times,* March 30, 2010, www.nytimes.com/2010/03/30 /business/global/30riotinto.html?pagewanted=all; James T. Areddy, "American Jailed in China Released, Arrives Home," *Wall Street Journal,* April 3, 2015, http://www.wsj.com/articles/china-said-to-be-deporting-u-s-geologist -jailed-on-spy-charges-1428094957.

5. Information Office of the State Council of the People's Republic of China, *Situation and Policies of China's Rare Earth Industry,* 1st ed. (Beijing, 2012), www.miit.gov.cn/n11293472/n11293832/n12771663/n14676956 .files/n14675980.pdf; China Radio International, "China Must Tackle Rare Earth Industry Chaos," August 9, 2014, english.cri.cn/12394/2014/08/09 /2743s839686.htm.

6. China US Focus, "Squaring the Circle: Rule According to Law in a One-Party State," October 30, 2014, www.chinausfocus.com/political -social-development/squaring-the-circle-rule-according-to-law-in-a-one -party-state/#sthash.9AUEM5uP.dpuf.

7. Chinese rare earth industry official, interview by David Abraham, Beijing, September 16, 2014; "Overseas Competition Forces Chinese Rare Earth Miners to Cut Glut," *Global Times,* December 23, 2013, www .globaltimes.cn/content/833308.shtml.

8. Chuin-Wei Yap, "China Plans To Create Iron-Ore Mining Giant," *Wall Street Journal,* March 20, 2014, online.wsj.com/articles/SB100 01424052702303802104579450302588753602; Wayne Arnold, "China's Global Mining Play Is Failing to Pan Out," *Wall Street Journal,* September 15,

2014 online.wsj.com/articles/chinas-global-mining-play-is-failing-to-pan
-out-1410402598.

9. "Hopes of Sparking Political Change Have Come to Nothing
So Far," *Economist,* December 10, 2011, www.economist.com/node/21541461.

10. Andrew Jacobs and Chris Buckley, "China Moves to Reinforce
Rule of Law, with Caveats," *New York Times,* October 24, 2014, www.nytimes
.com/2014/10/24/world/asia/china-moves-to-enact-rule-of-law-with
-caveats.html.

11. "China Denies Using Antimonopoly Law to Target Foreign Companies," *Wall Street Journal,* September 11, 2014, http://online.wsj.com
/articles/china-denies-using-antimonopoly-law-to-target-foreign
-companies-1410429955; China US Focus, "Squaring the Circle"; Jie Yang
and Laurie Burkitt, "China Defends Its Antimonopoly Probes," *Wall Street
Journal,* September 12, 2014, online.wsj.com/articles/SB3000142405297020
4168304580147352402004666. The U.S. Chamber of Commerce claimed
that for over six years Chinese authorities have been using the law to forward industrial policy goals. Other international organizations have lodged
similar complaints.

12. Zena Olijnyk, "Rare Earth: Canada's AMR Technologies Mines
Riches from China's Clay," *Canadian Business,* November 8, 2004, misc
.invest.stocks.narkive.com/uBHNu609/rare-earth-amr-tsx; "This Month in
Mining: China," *Engineering and Mining Journal* 200, no. 5 (2000), http://
www.highbeam.com/doc/1P3-58975170.html; "China Imposing Quotas on
Rare Earths Exports," *American Metal Market* 107 (February 1999), busi-
ness.highbeam.com/436402/article-1G1-53998245/china-imposing-quotas
-rare-earths-exports; "Chinese Rare-Earth Enterprises Meet to Streamline
Industry," *AsiaPulse News,* February 3, 1999, 1008034u0258; *Business Insights: Essentials,* accessed November 8, 2014, http://bi.galegroup.com
/essentials/article/GALE|A53695412/28calb8a9162fa4e234aa7df427082c7.

13. Li Jiabao and Liu Jie, "Rare Earth Industry Adjusts to Slow Market," *China Daily,* September 7, 2009, http://www.chinadaily.com.cn/bw
/2009-09/07/content_8660849.htm.

14. Helen Sun, "China Cuts Rare Earth Export Quota 72%, May Spark
Trade Dispute With U.S.," *Bloomberg,* July 9, 2010, http://www.bloomberg
.com/news/2010-07-09/china-reduces-rare-earth-export-quota-by-72-in
-second-half-lynas-says.html; V. Zepf, *Rare Earth Elements: A New Approach to the Nexus of Supply, Demand and Use* (Heidelberg: Springer, 2013).

15. Author's estimates are based on discussions with the China Society of Rare Earths, interview by David Abraham, Beijing, September 16,
2014.

16. Metal-Pages, "15 Antimony Smelters in China's Lengshuijiang to Merge," December 13, 2013, www.metal-pages.com/news/story/75928/15 -antimony-smelters-in-chinas-lengshuijiang-to-merge/. China plans to merge fifteen antimony smelters due to overcapacity concerns in the Lengshuijiang area, home to 60 percent of the country's antimony production in 2014. In 2003, China had four hundred producers of antimony, which is prized for its ability to create flame-resistant material as well as lead batteries. By 2011, the country had less than twenty producers. Judith Chegwidden and Jack Bedder, *Antimony: Changes in the Pattern of Supply and Demand* (Roskill Information Services, 2012), available at http://bit.ly /1IhN5j9. China.org, "Metals Firm Sees Shares Rise on Market Debut," October 10, 2012, www.china.org.cn/business/2012-10/10/content_26744718 .htm; Investor Intel, "China to Stockpile Heavy Rare Earths Again," October 17, 2012, investorintel.com/rare-earth-intel/china-to-stockpile-heavy -rare-earths-again/; China Tungsten Industry Association, "Tungsten Ore Supply Situation Is Difficult to Alleviate in Short-Term," accessed December 7, 2014, www.ctia.com.cn/TungstenNews/Print.asp?ArticleID=92789; Bob Davis, "China's State-Owned Sector Gets a New Boost," *Wall Street Journal,* February 23, 2014, online.wsj.com/articles/SB10001424052702303 63640457939693323203554 4. As China Nonferrous Mining notes in its 2012 prospectus, "we enjoy governmental support and preferential treatment in credit borrowing from banks and tax payments."

17. Marc Humphries, *China's Mineral Industry and U.S. Access to Strategic and Critical Minerals: Issues for Congress,* (CRS Report No. R43864), Washington, DC: Congressional Research Service, March 20, 2015, fas.org /sgp/crs/row/R43864.pdf.

18. Minor Metals Trade Association, "The Fanya Metal Exchange and the Impact of Conflict Mineral," July 2014, www.mmta.co.uk/newsletter /crucible.

19. Japanese government consultant, interview by David Abraham, Tokyo, August 4, 2013; Japanese government officials, interviews by David Abraham, Tokyo, 2012 and 2013.

20. Willem Thorbecke, "Exchange Rate Pass-Through in the Japanese Electronics Industry," *RIETI,* March 12, 2012, accessed December 7, 2014, http://www.rieti.go.jp/en/columns/a01_0338b.html.

21. "FIRB Approves Lynas Deal," *Sydney Morning Herald,* April 7, 2011, www.smh.com.au/business/firb-approves-lynas-deal-20110407-1d4x9.html.

22. Green Car Congress, "Report: METI, Toyota, Others to Ally to Develop Rare-Earth Recycling Tech," July 2012, www.greencarcongress .com/2012/07/ree-20120715.html.

23. Hiroshi Kawamoto and Wakana Tamaki, "Trends in Supply of Lithium Resources and Demand of the Resources for Automobiles," *Quarterly Review* 39, April (2011), www.nistep.go.jp/achiev/ftx/eng/stfc/stt039e/qr39pdf/STTqr3904.pdf. In World War I, Germany owned a third of Chile's production of nitrates to make explosives and fertilizers, but, because Germany had little control of the sea lanes it couldn't get the resources. Alfred E. Eckes, *The United States and the Global Struggle for Minerals* (Austin: University of Texas Press, 1979), 13.

24. David Abraham, "Geopolitics and Minor Metals," paper presented at the Minor Metals Metal-Pages Conference, Shanghai, China, September 2013.

25. European Commission, "Defining 'Critical' Raw Materials—Raw Materials—Enterprise and Industry," accessed December 7, 2014, ec.europa.eu/enterprise/policies/raw-materials/critical/index_en.htm.

26. Henrike Sievers, Bram Buijs, and Luis A. Tercero Espinoza, "Limits to the Critical Raw Materials Approach," *Proceedings of the ICE: Waste and Resource Management* 165, no. 4 (2012): 201–8, doi:10.1680/warm.12.00010, in Gus Gunn, *Critical Metals Handbook*, American Geophysical Union and Wiley, https://paperc.com/books/critical-metals-handbook/146308/A_9781118755211_c01. 2014.

27. U.S. Department of Energy, *Critical Materials Strategy*, 1st ed., e-book (2010), 90, available at http://energy.gov/sites/prod/files/edg/news/documents/criticalmaterialsstrategy.pdf. "[T]he market dynamics that affect . . . key materials vital to the commercialization of clean energy technologies are not captured by traditional economic models or simple economic analyses."

28. World Energy Outlook for 2002 and 2004, Worldenergyoutlook.org. The World Energy Outlook for 1998 uses Mtce, which was converted into Mtoe using the IEA's own converter. See "IEA: Unit Converter," http://www.iea.org/stats/unit.asp; Maria van der Hoeven, *Coal Medium-Term Market Report 2013*, www.iea.org/Textbase/npsum/MTCoalMR2013SUM.pdf. China used 2,806 Mtce in 2012 or 7,500 metric tons. Hartmuth Zeiß, *Global Coal: Trends and Outlook: What It Means for the EU?* Thirteenth European Round Table on Coal, March 17, 2011; American Physical Society, "New APS-MRS Report: Energy Critical Elements—Developing New Technologies to Foster U.S. Energy Independence," 2011, www.aps.org/about/pressreleases/elementsreport.cfm.

29. Robert Jaffe, interview by David Abraham, Cambridge, MA, February 21, 2013.

30. Statistic from National Marine Fisheries Service. Paul Greenberg, "Why Are We Importing Our Own Fish?" *New York Times,* June 20, 2014, nyti.ms/1pn57Fw.

31. David Kramer, "Rare-Earth Metals Shortage Eases, for Now," *Physics Today,* 2011, scitation.aip.org/content/aip/magazine/physics today/news/10.1063/PT.4.0573, doi:10.1063/pt.4.0573; "Testimony of Dr. Robert Jaffe," before the Energy and Environment Subcommittee of the House Science Committee, 2011, https://science.house.gov/sites/republicans.scie nce.house.gov/files/documents/hearings/120711_Jaffe.pdf.

32. Kramer, "Rare-Earth Metals Shortage Eases."

33. As the American geologist Josiah Spurr said in 1920, "Our own vast mineral wealth is so abundant that not till recently has American capital and enterprise found it necessary to adventure into the outside world, as European nations have done long ago." Spurr, *Political and Commercial Geology and the World's Mineral Resources* (New York: McGraw-Hill, 1920), vi.

34. Joe Stephens and Carol D. Leonnig, "Solyndra: Politics Infused Obama Energy Programs," *Washington Post,* December 14, 2010, www .washingtonpost.com/solyndra-politics-infused-obama-energy-programs /2011/12/14/gIQA4HllHP_story.html.

35. Jeff Phillips, telephone interview by David Abraham, November 8, 2013.

36. In 1994, ninety commodities were stockpiled. Fifteen years later, in 2009, there were twenty-four because the United States had sold off its supplies. Now, in 2014, the United States is increasing its stockpile but with the current cutbacks in spending, the U.S. government is unlikely to commit billions more to stockpile resources. Committee on Armed Services, House of Representatives, *Hearing: Proposed Reconfiguration of the National Defense Stockpile,* 1st ed., e-book (Washington, DC, 2009), http://www.hsdl .org/?view&doc=112062&coll=limited.

37. Brian J. Fifarek, Francisco M. Veloso, and Cliff I. Davidson, "Offshoring Technology Innovation: A Case Study of Rare-Earth Technology," *Journal of Operations Management* 26, no. 2 (2008): 222–38, doi: 10.1016/j.jom.2007.02.013.

38. Statistics from U.S. Patent Office, http://patft.uspto.gov.

39. Manuel Quinones, "Push to Rebuild Depleted U.S. Workforce Begins in the Classroom," February 13, 2012, E&E, www.eenews.net/stories /1059959839.

40. Center for Strategic and International Studies Nuclear Energy Program, *Restoring U.S. Leadership In Nuclear Energy: A National Security*

Imperative (Lanham, MD: Roman and Littlefield, 2013), available at http://
csis.org/files/publication/130614_RestoringUSLeadershipNuclearEnergy
_WEB.pdf. The nuclear accident at Three Mile Island in 1978 is seen as the
catalyst for change against nuclear development in the United States.

41. Juliette Garside, "Apple Creates 2,000 Jobs Shifting Production
Back to US," *Guardian*, November 5, 2013, http://www.theguardian.com
/technology/2013/nov/05/apple-creates-us-jobs-renewable-energy; Mark
Crawford, "Manufacturing in America: Bigger, Better and Bolder," *Area Development*, Q1 2014, www.areadevelopment.com/manufacturing-industrial
/Q1-2014/US-manufacturing-upswing-innovation-productivity-11278770
.shtml; Boston Consulting Group, "U.S. Executives Remain Bullish on
American Manufacturing, Study Finds," 2014, www.bcg.com/media/press
releasedetails.aspx?id=tcm:12-174453; Antoine van Agtmael and Fred Bakker, "Made in the U.S.A. (Again)," *Foreign Policy*, March 28, 2014, foreign
policy.com/2014/03/28/made-in-the-u-s-a-again/.

11
How to Prosper in the Rare Metal Age

1. "Definition of: Centronics Interface," *PC Magazine*, accessed December 19, 2014, www.pcmag.com/encyclopedia_term/0,1237,t=Centronics
interface&i=57167,00.asp; Michelle Kessler, "First Internet Message Sent
40 Years Ago Today," *USA Today*, October 29, 2009, accessed December 19,
2014, content.usatoday.com/communities/technologylive/post/2009/10/620
000700/1#.T_JppvXueSo. The journal *Administrative Management* predicted
that "by the end of the 1970s, we should have climbed out of the Gutenberg
rut," which would have a profound influence on resource use. "Management
and the Information Revolution," *Administrative Management* 31, no. 1 (January 1970), quoted in Gil Press, "The Paperless Office of the Future, Still," The
Story of Information, 2011, http://infostory.com/2011/09/06/the-paperless
-office-of-the-future-still/. Five years later, the term "paperless office" first appeared in *BusinessWeek*, summing up the belief that paper's days were numbered. "The Office of the Future," *Bloomberg Business Week*, June 30, 1975,
accessed December 19, 2014, www.businessweek.com/stories/1975-06-30/the
-office-of-the-futurebusinessweek-business-news-stock-market-and
-financial-advice. In the same article, Vincent E. Giuliano, of the consultancy
Arthur D. Little, echoed the journal's sentiments when he said that paper use
in the business world would decline through the 1980s, "and by 1990, most
record-handling will be electronic."

2. United Nations Statistics Division, "United Nations Statistics Database," accessed December 19, 2014, http://unstats.un.org/unsd/databases .htm; Michael Saylor, *The Mobile Wave: How Mobile Intelligence Will Change Everything* (New York: Vanguard Press, 2012). By 2020 that number is set to grow by 20 percent to 500 million tons. Forest Industries, "Global Paper Consumption Is Growing: Paper and Pulp Industries—Paper, Paperboard and Converted Products," July 10, 2013, accessed December 19, 2014, www .forestindustries.fi/industry/paper_cardboard_converted/paper_pulp /Global-paper-consumption-is-growing-1287.html.

3. Ernst Ulrich von Weizsäcker et al., "Decoupling 2 Technologies, Opportunities and Policy Options," United Nations Environment Programme, 2014, http://www.unep.org/resourcepanel/Portals/24102/PDFs /IRP_DECOUPLING_2_REPORT.pdf; "Growth in Global Materials Use, GDP and Population during the 20th Century: Online Global Materials Extraction 1900–2009 (update 2011)," Social Ecology Vienna, 2009, accessed December 19, 2014, uni-klu.ac.at/socec/inhalt/3133.htm; von Weizsäcker et al., "Decoupling 2 Technologies."

4. *Deloitte Digital Democracy Survey*, 8th ed. (e-book, 2014), http:// www2.deloitte.com/content/dam/Deloitte/us/Documents/technology -media-telecommunications/us-tmt-deloitte-digitaldemocracy_102014 .pdf.

5. Laura Hubbard, Consumer Electronics Association, e-mail, January 31, 2014.

6. The growth rate from 2011 to 2015 was estimated to be 17.2 percent. "Global E-Waste Management Market (2011–2016)," Markets and Markets, 2011, accessed December 19, 2014, www.marketsandmarkets.com/Market -Reports/electronic-waste-management-market-373.html; U.S. Environmental Protection Agency, *Municipal Solid Waste Generation, Recycling, and Disposal in the United States: Facts and Figures for 2012*, http://www .epa.gov/waste/nonhaz/municipal/pubs/2012_msw_fs.pdf; von Weizsäcker et al., "Decoupling 2 Technologies."

7. "Toxic Computer Waste in the Developing World," *ScienceDaily*, June 3, 2014, accessed December 19, 2014, http://www.sciencedaily.com /releases/2014/06/140603114327.htm; Andy Walton, "Life Expectancy of a Smartphone," *Small Business*, accessed December 19, 2014, http://small business.chron.com/life-expectancy-smartphone-62979.html. A smartphone's average life span is twenty-two months. David Pogue, "Should You Upgrade Your Phone Every Year?—Not Anymore," *Scientific American*, August 20, 2013, accessed December 19, 2014, www.scientificamerican .com/article/should-you-upgrade-your-phone-every-year-not-anymore/.

8. "Apple's Latest 'Innovation' Is Turning Planned Obsolescence into Planned Failure," iFixit Blog, January 20, 2011, accessed December 19, 2014, www.ifixit.com/blog/2011/01/20/apples-latest-innovation-is-turning -planned-obsolescence-into-planned-failure/; Apple, "iPhone Support— Screen Damage," iPhone Screen Damage Repair, accessed May 19, 2014, https://www.apple.com/support/iphone/repair/screen-damage/; Brian Clark Howard, "Planned Obsolescence: 8 Products Designed to Fail," *Popular Mechanics,* accessed December 19, 2014, www.popularmechanics.com/tech nology/planned-obsolescence-460210#slide-8. T-Mobile plans allow phone upgrades every six months as of January 2014. Sprint ended its program for annual upgrades in January 2014. T-Mobile, "iPhone 6 Is Here: Phone Upgrade Anytime," accessed May 7, 2014, t-mo.co/1sMUCCi.

9. Tatsuo Ota, Lecture at the Metal Bulletin Events Asian Ferro-alloys Conference, "Trade and overseas investments in the chrome, nickel and manganese markets from Japan," March 25–27, 2009; unnamed Mitsubishi corporate employee, interview by David Abraham, Tokyo, Japan, November 9, 2011.

10. Malik Crawford and Jonathan Church, eds., "CPI Detailed Report Data for January 2014," Bureau of Labor Statistics, http://www.bls.gov /cpi/cpid1401.pdf.

11. Rob Tenent, interview by David Abraham, Golden, CO, July 13, 2013; Devin Powell, "Smart Glass Blocks Heat or Light at Flick of a Switch," *Nature,* August 14, 2013, www.nature.com/news/smart-glass-blocks-heat-or -light-at-flick-of-a-switch-1.13558; Tenent, interview, July 13, 2013.

12. National Renewable Energy Laboratory, "Research Support Facility," January 9, 2014, www.nrel.gov/sustainable_nrel/rsf.html; GreenBiz, "NREL Opens State-of-the-Art Net-Zero Energy Facility," July 8, 2010, www .greenbiz.com/news/2010/07/08/nrel-opens-state-art-net-zero-energy -facility; Powell, "Smart Glass Blocks Heat;" Michael Silver, telephone interview by David Abraham, March 13, 2014.

13. David Szondy, "GE Reveals Vision for Homes of the Not-Too-Distant Future," October 13, 2013, www.gizmag.com/ge-future-home-2025 /29282; General Electric, "Home 2025: GE Envisions Home of the Future," http://www.geappliances.com/home2025/.

14. "Material requirements per unit generation for low-carbon technologies can be higher than for conventional fossil generation: 11–40 times more copper for photovoltaic systems and 6–14 times more iron for wind power plants." E. Hertwich, E. van der Voet, S. Suh, A. Tukker, M. Huijbregts, P. Kazmierczyk, M. Lenzen, J. McNeely, and Y. Moriguchi, "Assessing the Environmental Impacts of Consumption and Production: Priority

Products and Materials," A report of the Working Group on the Environmental Impacts of Products and Materials to the International Resource Panel, United Nations Environment Programme, 2010, http://www.green ingtheblue.org/sites/default/files/Assessing%20the%20environmental%20 impacts%20of%20consumption%20and%20production.pdf; John Heggestuen, "One in Every 5 People in the World Own a Smartphone, One in Every 17 Own a Tablet," *Business Insider,* December 15, 2013, www.businessinsider .com/smartphone-and-tablet-penetration-2013-10; Pew Research Center Internet American Life Project, "Device Ownership over Time," November 13, 2013, www.pewinternet.org/data-trend/mobile/device-ownership/; Dave Evans, "The Internet of Things: How the Next Evolution of the Internet Is Changing Everything," CISCO, April 2011, http://www.cisco.com /web/about/ac79/docs/innov/IoT_IBSG_0411FINAL.pdf.

15. Nicola Twilley, "What Do Chinese Dumplings Have to Do with Global Warming?" *New York Times,* July 26, 2014, www.nytimes.com/2014 /07/27/magazine/what-do-chinese-dumplings-have-to-do-with-global -warming.html.

16. National Intelligence Council, "Global Trends 2030: Alternative Worlds," December 1, 2012, accessed December 19, 2014, www.dni.gov/index .php/about/organization/global-trends-2030; Kohmei Halada, Masanori Shimada, and Kiyoshi Ijima, "Forecasting the Consumption of Metals up to 2050," *Journal of the Japan Institute of Metals* 71, no. 10 (2007): 831–39.

17. Roger Agnelli, interview by David Abraham, São Paulo, Brazil, May 6, 2013.

18. Gordon Gable, "Top 10 Bad Tech Predictions," *Digital Trends,* November 4, 2012, accessed December 19, 2014, http://www.digitaltrends.com /features/top-10-bad-tech-predictions/4/#ixzz2s4mjo7iC.

19. Erin Skarda, "Top 10 Failed Predictions," *Time,* October 21, 2011, accessed December 20, 2014, http://content.time.com/time/specials/pack ages/article/0,28804,2097462_2097456_2097467,00.html.

20. Bob Metcalfe, "From the Ether: Predicting the Internet's Catastrophic Collapse and Ghost Sites Galore in 1996," *Info World,* December 4, 1995, http://bit.ly/1AlCm3B.

21. K. Binnemans, P. T. Jones, K. Acker, B. Blanpain, B. Mishra, and D. Apelian, "Rare-Earth Economics: The Balance Problem," *JOM,* 2013, http://www.kuleuven.rare3.eu/wp-content/plugins/rare/images/papers /binnemans_jom_2013.pdf; Molycorp, "Advanced Communications," accessed December 19, 2014, www.molycorp.com/technology/green-element -technologies/advanced-communications; APS Physics, "New Prototype Magnetic Refrigerators Hold Commercial Promise," accessed December

19, 2014, www.aps.org/publications/apsnews/200305/refrigerators.cfm; Binnemans et al., "Rare-Earth Economics: The Balance Problem."

22. American Chemical Society, "Toward Lowering Titanium's Cost and Environmental Footprint for Lightweight Products," December 18, 2013, accessed December 19, 2014, http://www.acs.org/content/acs/en/pressroom/presspacs/2013/acs-presspac-december-18-2013/toward-lowering-titaniums-cost-and-environmental-footprint-for-lightweight-products.html; America's Navy, "Future Naval Force May Sail with the Strength of Titanium," April 3, 2012, http://www.navy.mil/submit/display.asp?story_id=66264; Minor Metals Trade Association, "Minor Metals in the Periodic Table," accessed December 19, 2014, http://www.mmta.co.uk/metals/Ti.

23. The Colorado School of Mines received funding from the U.S. government after the government set up a Critical Materials Hub two years ago.

24. Klaus Jaffe, Mario Caicedo, Marcos Manzanares, Mario Gil, Alfredo Rios, Astrid Florez, Claudia Montoreano, Vicente Davila, and Alejandro Raul Hernandez Montoya, "Productivity in Physical and Chemical Science Predicts the Future Economic Growth of Developing Countries Better than Other Popular Indices," *PLoS ONE* 8, no. 6 (2013): E66239, available at www.plosone.org/article/info%3Adoi%2F10.1371%2Fjournal.pone.0066239#pone-0066239-g001; Materials Research Society, "Research Investment: Economic Growth—Research Materials Innovations," www.mrs.org/return-on-investment-economic-growth.

25. According to the National Science Foundation 743 doctorates were awarded in 2012. National Science Foundation, "Science and Engineering Doctorates," November 2014, http://www.nsf.gov/statistics/sed/2012/data_table.cfm.

26. "Big Demands and High Expectations: The Deloitte Millennial Survey," January 2014, http://www2.deloitte.com/al/en/pages/about-deloitte/articles/2014-millennial-survey-positive-impact.html. An example of breakthroughs that could lessen dependence on rare metals includes perovskites use in solar applications. Mark Peplow, "Perovskite Solar Cell Bests Bugbears, Reaches Record Efficiency," IEEE Spectrum, January 7, 2015, http://spectrum.ieee.org/energywise/green-tech/solar/perovskite-solar-cell-bests-bugbears-reaches-record-efficiency.

27. Walter R. Stahel, "Caterpillar Remanufactured Products Group," Product-Life Institute, accessed December 19, 2014, http://www.product-life.org/en/archive/case-studies/caterpillar-remanufactured-products-group; Caterpillar, "Growth from Sustainability: Caterpillar's Experiences plus

Wishes for Collaboration with Singapore-based Companies," accessed December 19, 2014, http://www.simtech.a-star.edu.sg/SMC/media/1198 /growth_from_sustainability-caterpillarsexperiencesnwishesforcollabora tionwithsingaporebasedcompanies.pdf; Caterpillar, "2012 Year in Review: Solid Rock," 2013, http://s7d2.scene7.com/is/content/Caterpillar/C10005383.

28. Christian Hagelüken and Christina E. M. Meskers, "Complex Life Cycles of Precious and Special Metals," in *Linkages of Sustainability*, ed. T. E. Graedel and E. van der Voet, 165–197 (Cambridge, MA: MIT Press, 2010), published online August 2013, http://oxfordindex.oup.com/view/10.7551 /mitpress/9780262013581.003.0010?rskey=Hwg0Ci&result=2.

29. The U.S. EPA estimated in 2011 that the country threw away 2.4 million tons of electronic products. Environmental Protection Agency (EPA), "Statistics on the Management of Used and End-of-Life Electronics," accessed December 19, 2014, http://www.epa.gov/epawaste/conserve /materials/ecycling/manage.htm. Electronics TakeBack Coalition, "E-Waste Problem Overview."

30. "Why Are Rare Earth Metals So Important?" accessed December 19, 2014, lamprecycling.veoliaes.com/newsletter/September2013/6.

31. Desiree Mohindra, "Circular Economy Can Generate US$ 1 Trillion Annually by 2025," World Economic Forum, accessed December 19, 2014, www.weforum.org/news/circular-economy-can-generate-us-1-trillion -annually-2025.

32. Francois-Xavier Lienhart, "The Implementation of an Energy-Saving Society Contributes to the Environment, People and Economy," presentation, Ministry of Economy, Trade and Industry, Tokyo, Japan, November 1, 2011.

33. According to Elias Strangas, head of Michigan State's Electrical Machines and Drives Laboratory, if it were not for the events of 2010, the permanent magnet would be used more widely. Statistics from Steve Constantinides show a drastic drop in magnet use after 2010. Steve Constantinides, "Magnetic Materials and Market Sit-Rep," presentation, The Motor & Motion Association, November 4–6, 2014.

34. International Energy Agency, accessed December 19, 2014, http:// www.iea.org/.

35. Thomas Jefferson National Accelerator Facility, Office of Science Education, "The Element Terbium: It's Elemental," accessed December 19, 2014, http://education.jlab.org/itselemental/ele065.html.

36. Masato Sagawa, e-mail, December 8, 2014.

Acknowledgments

The idea for a book on rare metals started when *The Economist* data guru, Kenneth Cukier, asked for my thoughts on rare earth metals over wine in Tokyo one evening in October 2010. His questions were important to my initial decision to write. But without the enthusiasm of Gillian MacKenzie, my agent and encourager-in-chief who urged me to write and supported me along the way, nothing would have materialized. Eli Kintisch and Maryann Matthews, who never ceased to find ways to improve my writing, were crucial in shaping the piece, but no shaping was more critical than that of my first line editor, Barbara.

A special thanks to Joe Calamia at Yale University Press, who thought the world was ready to know about these rare metals and who constantly pushed me for better explanations of the science. He helped transform an idea into a manuscript and then a book along with his incredibly patient team, Annie Imbornoni, Samantha Ostrowski, and Nancy Ovedovitz. I'm indebted to those who took time out of their days to show me around their offices, countries, and towns and those who spent time answering the most basic questions over and over again:

Elisa Alonso, Ed Becker, Karl Gerald van den Boogaart, Zhangheng Chen, Steve Constantinides, Ken Deckinger,

Roderick Eggert, Eize de Vries, Thomas Graedel, Gareth Hatch, Zhu Hongmin, Robert Jaffe, Yujia He, Randy Kirchain, Alain Leveque, Anthony Mariano Sr., Shinsuke Murakami, Kevin Moore, Yuji Nishikawa, David O'Brock, Toru Okabe, Barbara Reck, Lisa Reisman, Michael Silver, John Sykes, Alex and Dan Teran, Stan Trout, and the government offices of Araxá, Brazil; Sillamäe, Estonia; and Bangka, Indonesia. Special thanks also to John Smith and Dominic Boyle of 5NPlus, the incredible and generous faculty and students at the Colorado School of Mines, notably Patrick Taylor, Ron Cohen, Corby Anderson, Caelen Anderson, and Joseph Grogan, and the folks at the National Renewable Energy Laboratory including Michael Woodhouse and Robert Tenent, as well as Alia Adistya Nasier for the support and needed translation.

A number of others spent time on the phone and generously offered their opinions and contacts, including Noah and Danny Lehrman, Maria Cox of the Minor Metals Trade Association, Judith Chegwidden of Roskill, Claire Miko of Indium Corporation, Tracy Weslosky of InvestorIntel, Jack Lifton and Gal Luft, analysts at the U.S. Geological Survey, and Jeff Green.

I'm grateful to the hundreds of other people who answered questions to help me deepen my understanding of the science, trading, and use of rare metals.

I am also very thankful for the insight of those in the trenches who added their deep insights to the book: Luisa Moreno, William H. Hess, Kenn Cukier, General Robert Latiff, Nigel Tunna of Metal-Pages, and Dudley Kingsnorth. Also, to those who spent hours reading through drafts—Lionel Beehner, Adam G. Hinds, Andrew Huszar, Meredith Ludlow, Michael Shane, Cleo Sonnenborn, Lizzie Wade—I'm indebted. I extend a special thanks to the research institutions that supported my work, the Council on Foreign Relations,

Institute for the Analysis of Global Security, and Tokyo University, and for the generous support of the Hitachi Corporation and employees including Minoru Tsukada, Masanori Ueda, and Yoko Yamazaki. The Research Institute of Economy, Trade and Industry (Japan) and Yasuhiko Yoshida and Junichiro Kuroda were particularly helpful in ensuring that the ministry staff was available to answer questions. The views expressed in the previous pages are mine and in no way reflect the views of any organization that I have been a part of. To be sure, without the support of Brian Dusza and his gracious accommodations in Jakarta, and likewise Denis Chichkine in Tokyo, the writing of this book would have been far more costly. Many thanks also to Swares Sinaga who helped in research during the early stages. And a special note of appreciation to those who took time to answer questions but wanted anonymity.

Finally to Mike, Joan, Eden, and Meyer, well, just because.

Index

A-12 spy plane ("Oxcart," "Titanium Goose"), 155–57, 158–59
Abate, Victor, 139
Abel, Bas van, 110
Abraham, David S., 25–27
Acer, 122
Acids: pollution from, 174, 176, 179–80; use in rare metals production, 44, 70, 71–72, 74, 76, 81–82; use of term, xiii
Adams, John, 167
Administrative Management (journal), on Gutenberg rut, 292n1
Advanced Materials Japan, 90
Advanced metals. *See* Rare metals
Advanced Research Projects Agency Network (ARPANET), 214
Aerospace industry, 45, 163, 263n30
Agnelli, Roger, 219
Airbus, 128
Aircraft Industries of America, 274n6

Airplanes, 128–31, 155–60, 168, 274n6, 279n33
Air pollution, in China, 153, 281n2
Alcorn, Walter, 119
Alonso, Elisa, 136, 222
Alstom, on wind turbines, 137, 266n7
Aluminum, 78, 81–82, 143, 155, 164, 169, 190, 285n34
American Chemical Society, on rare metals shortages, 12
American Elements, 218
American Metals, 161
American Physical Society, 135, 208
American Vanadium, 52
Anderson, Caelen, 86, 87
Anderson, Corby, 85
Anemia, treatment for, 278n28
Animated dolls, 119–20
Antimony, 4, 40, 103, 205, 207, 240n34, 289n16
Apple, 1, 10, 110, 178, 212, 223
Araxá, Brazil, tourism and mining in, 38–40
ARPANET (Advanced Research Projects Agency Network), 214

Arrhenius Carl Axel, 72
Asia, geopolitics of rare metals in, 195. *See also names of individual countries*
Assyrians (ancient), weaponry, 157–58
AT&T, iPhone sales, 1
Auer von Welsbach, Carl, 72
Australia, rare earth processing in, 183, 205
Automobiles, 141–48
Avalon Rare Metal, 54–56, 57, 63–64, 85

Ballmer, Steve, 1
Bangka, Indonesia, tin production in, 48, 105–6
Banks, investments in rare metals trading, 96
Baotou, China, 78, 176–77
Barium, 121
Base metals, 4, 29, 78, 101
Base Resources (Australia), 48
BASF, 3, 232n6
Battelle Memorial Institute, 274n6
Batteries, 116, 147–48, 188–89
Bayan Obo mine, 78, 175–76, 196
Becker, Ed, 142, 143
Belitung, Indonesia, tin production in, 48
Beryllium, 55, 121, 168, 260n14
Berzelius, Jöns J., 72
BHP Billiton, 58
Big Bertha gun, 160–61, 275n12
Big Data (Cukier and Mayer-Schönberger), 119
Bissel, Richard, 159
Bloomberg News: on CBMM, 42; on Colombian tungsten trade, 109

Boeing, 113, 128, 130–31
Boiridy, Mia, 85
Bombs, from airplanes, 279n33
Boogaart, Gerald van den, 33–35
Boron, 21, 26, 116, 121
Boston Consulting Group, 212
Boyle, Dominic, 163
Bre-X (exploration company), 59
Britain: export bans during WWI, 162–63; tungsten, actions on during WWII, 239n28
British Geological Survey, on Chinese production of critical materials, 236–37n18
Bronze, 157
Bronze Age, 12, 157, 274n7
Broxo company, 115
Bubar, Don, 55, 64
Bukit Merah, Malaysia, pollution in, 183
Burns, Stuart, 147
Business models, need for change in, 223–25
By-product production, 79–80

Cadmium, 3, 116, 148, 159, 167, 181, 258n3
Cadmium-tellurium thin films, 148–49
Calculators, 118–19
Canada: indium sales via telemarketing, 251n7; mining workforce, age of, 85
Carbon emissions, 152–53, 266n5, 281n2
Carnegie Mellon University, 211
Carneiro, Tadeu: on CBMM, 43, 46, 64–65; lack of investment worries, 54, 64; on niobium, 44;

as spokesperson for CBMM, 41;
on sustainability, 152, 153
Cars, 141–48
Cassiterites (tin ore), 105–6
Castilloux, Ryan, 116
Catalytic converters, 144–45
Caterpillar, 212, 223–24
CBMM (Companhia Brasileira de
Metalurgia e Mineração),
39–46, 54, 62, 64–66, 152–53,
242n6
Central Intelligence Agency
(CIA), 158
Centronics, 214
Ceramics, in wireless networks,
124
Cerium, 2, 35, 74, 75, 104, 140–41
CERN, Large Hadron Collider, 81
CFLs (compact fluorescent
lightbulbs), 150
Characteristics of rare metals, 3–4
Chicago Board of Trade, 101
Chicago Mercantile Exchange, 101
Chile, ore grade of lithium mines,
285n33
China: antimony production in,
289n16; CBMM ownership in,
42; coal demand in, 208; critical
material production, 236–
37n18; defense expenditures,
278n31; environmental issues,
153–54, 173–77, 281n2; export
ban on rare earth, x, 212; Hong
Kong, relationship with, 102;
Japan, conflict with, x, 15,
22–25, 165; Jiangxi, ore
processing in, 77, 82–85;
low-energy lighting production
in, 152; material production
costs, 240n33; rare earth

elements supply chain, control
of, 32–37; rare earth permanent
magnets in, 137; rare metal
exchanges, 96–98; rare metals
industry in, 194–200; refining
in, 75, 82–85; regulatory
environment, 99–101, 103–5,
202, 240n34, 288n11; steel
demand in, 11; technology use
in, 218; tungsten production in,
289n16; WTO membership of,
200–203
China Securities Regulatory
Commission, 99, 101
Chinese Society of Rare Earths,
176
CIA (Central Intelligence
Agency), 158
Circular economies, 225
Cisco, 218
Clean energy technologies,
290n27
Cloud storage, 122
CO_2 emissions, 152–53, 266n5,
281n2
Coal, 149–50, 178, 207, 208
Cobalt, 3, 18–21, 25, 28, 78, 101, 121,
128, 147, 219, 235nn5–6, 260n15
Cohen, Ronald R., 179–80, 184
Colombia: mineral trading as
funding for conflicts in, 109;
tungsten production in, 48
Colorado School of Mines, 79,
86–87, 296n23
Committee on Natural Resources
(U.S. House of Representatives),
210
Commodities, 91, 216, 220, 291n36
Compact fluorescent lightbulbs
(CFLs), 150

Companhia Brasileira de
Metalurgia e Mineração
(CBMM), 39–46, 54, 62, 64–66,
152–53, 242n6
Compaq, 255n35
Conferences on rare metals,
194–95
Conflicts, funding of, from rare
metals production, 108–12
Congo: conflict minerals from,
108–9, 111. *See also* Democratic
Republic of Congo; Zaire
Congress, materials report (1985),
239n30
ConocoPhillips, 86
Constantinides, Steve, 186
Consumer habits, 223
Copper: automobile engines and,
143; conductive powers of, 164;
demand for, 4, 215, 294n14;
Germany and, 29; grades of, 55,
79, 182; locations and availabil-
ity of, 157, 246n40; processing
of, 4, 76, 78–79, 177; recycling,
189; rhenium as by-product of,
128
Corporate issues, 38–66; Avalon
Rare Metal, 54–56; CBBM,
39–46, 64–66; commodity
investors, 60–62; junior mining
companies, 59–60; market mining, 56–59; rare metals mining,
investments in, 49–54; time to
production, 62–64; Tiomin
Resources, 46–48
Council on Foreign Relations, x
Cox, Maria, 92
Cree (lighting manufacturer), 151
Critical material, xiii. *See also*
Rare metals

Critical Materials Strategy
(Department of Energy), 207
Cukier, Kenneth, 119
Currid, Arch, 255n35

Da Costa, Jeová Moreira, 42
Daido Steel, 113
Dalahai, China, pollution in,
175–77
DDG 51 Aegis destroyers, 168
Decision-making processes, for
rare metal usage, 227–28
Deckinger, Ken, 187–88
Defense Logistics Agency, 240n31
Defense sector. *See* Military
(U.S.); Wars
Dell Corporation, 14, 224
Democratic Party of Japan, 23–24
Democratic Republic of Congo,
rare metals production in, 48
Deng Xiaoping, 32
Dentistry, historical origins, 115
Department of ___. *See* U.S.
Department
Design Journal, on Sinclair's
calculator, 118
Developing countries: challenges
to metal operations in, 48–49;
rise in standard of living in,
10–11; technological improve-
ments in, 218–19; technology
use in, 125–27
Diamond, Jared, 10
Diaoyu Islands (Senkaku Islands),
territorial dispute over, 22–24
Didymium, 72
Dingnan, China, 173–75
Diodes, 117, 164–65, 277n27,
278n30. *See also* Light-emitting
diodes (LEDs)

Dokai Bay, Japan, pollution in, 181
Dolls, animated, 119–20
Doping agents, in optical fibers, 262n23
Drive trains, for wind turbines, 137–39, 266n8, 268n14
Duclos, Steve, 129, 130, 132, 133
Dysprosium: demand for, 9, 22, 133, 220; in electric tooth-brushes, 116; lack-of-substitute risk, 207; in magnets, 4, 21–22, 25, 27–28, 205, 229, 267n9; potential scarcity of and reduced reliance on, 11–13, 139, 140, 229; properties of, 4; for renewable energy, 136, 138; sources of, 174. *See also* Neodymium-dysprosium magnets

East China Sea, territorial dispute in, 22–24
EcoATM, 188
Eco-recycling, 186–87
Egyptians (ancient), use of toothpaste, 257n1
Eiffel, Gustave, 44
Electric toothbrushes, 115–17, 258n3
Electric vehicles (EV), 136, 145–48, 270n30
Electronic waste, xi, 191–92, 216
Element isolation, 72–76
Elevators, energy use of, 150
Ellen MacArthur Foundation, 225
Energy: carbon emissions and production of, 266n5; clean energy technologies, 290n27; conservation of, 225; renewable energy, 135, 136–37; use in mining, 283n19

Energy Independence and Security Act (2007), 151
Environmental needs, 134–54; automobiles and, 141–44; cerium magnets and, 140–41; in China, 153–54; CO_2 emissions, 152–53; conclusions on, 154; electric vehicles, 145–48; lighting, 150–52; overview, 134–36; platinum group metals and, 144–45; rare earth crisis, effects of, 138–40; renewable energy, need for, 135, 136–37; solar panels, 148–50; wind turbines, 137–38. *See also* Sustainable use
Environmental Protection Agency, 187
Environmental regulations, 184
Erceg, Luka, 211–12
Ernst and Young, on cost overruns, 58
Estonia, energy production in, 134–35
European Union: on Chinese export controls, 36; conflict materials, actions on, 111; rare metal security strategy, 205; on rare metals shortages, 136–37
Europium, 2, 151, 167
EV (electric vehicles), 136, 145–48, 270n30
Extractive Metallurgy of Rare Earths (Krishnamurthy and Gupta), 70
Eyang Subur, 126

F-35 Lightning II aircraft, 168
Facebook, 126

Factory Number 7 (Silmet), 67–69, 72–76, 80–82
Fairphone, 110
Fanya Metal Exchange, 97
FARC rebels (Colombia), 109
Fiber optics, 124
Fink, Colin, 162
First Solar, 149
Flat-screen technology, 122–24
Fleming, Rowland, 59
Flotation (processing method), 162
Ford, Henry, 52
Ford Motor Company, 132, 144–45
Fraud, 59, 98
Free markets, U.S. faith in, 209
Frontier Rare Earth, 76–77
Fujitsu, 20
Fundraising, for metals production, 245n29
Futures trading, 100–101

Gadgets: electronic, 119–20; increasing use of, 215; sustainability and, 223; use of resources for, 9–10
Gadolin, Johann, 72
Gadolinite (yetterbite), 72
Gadolinium, 221
Gallium, 3, 78, 118, 121, 149, 168, 220–21
"Gambler" investor type, 60–61
Gan Yong, 33
Gearboxes, of wind turbines, 137, 266n8
General Electric (GE), 35, 114, 129–31, 139, 151, 212, 218
General Motors (GM), 141, 142, 144, 235n5

Geopolitics, 194–213; China, control of rare metal industry in, 197–200; China, importance of rare metals industry to, 194–97; China, WTO membership of, 200–203; conclusions on, 212–13; European Union, rare metal security strategy, 205; future resource crunches and, 219; impact of, 139–40; Japan, rare metal security strategy, 203–5; resource constraints and, 206–8; U.S., rare metal security strategy, 205, 208–12. See also National struggles
Geothermal energy as field of study, 212
"German dentist" investor type, 60
Germanium: germanium tetrachloride ($GeCl_4$), 124, 262n23; in integrated circuits, 117; production of, 278n30; resource demands for, 179, 209; source of, 76; use before WWII, 278n28; wars, use in, 163–66
Germany: metallurgical capacity, 29; tungsten, actions on during WWII, 239n28; WWI weaponry, 160–61
Ghana, recycling in, 191
Glencore (commodity trader), 103
"Glenn Beck devotee" investor type, 60
Global forum, need for, 228–29
Gold, xiii, 4, 59, 60, 69, 76, 94, 108, 158, 187, 190, 191, 215
Government Accountability Office, 169

Governments: research spending by, 221–22; role in Rare Metal Age, 225–27, 228. *See also names of individual countries*
Graedel, Thomas, 126, 169, 191, 234n19
Graphite, 147
Greenhouse gases, 178
Green technologies, 12, 135, 179. *See also* Sustainable use
Gschneidner, Karl, 28, 130
Guliano, Vincent E., 292n1
Gussack, David, 185
Gutenberg rut, 292n1

Habord, James, 29
Haig, Alexander, 19
Halada, Kohmei, 177–78, 179
Halliburton, 86–87
Hamano, Masaaki, 21
Hastings, Richard Norman ("Doc"), 210
Hatch, Gareth, 138, 147
Heavy metal, 175
Heavy rare earths, 57, 75, 194, 205
Hess Corporation, 86–87
High-performance materials, need for, 169, 171–72
High-tech products, 179, 215
High-tech supply chain, 33
Hiranuma, Hikaru, 187
Hitachi Corporation, 186–87, 189–90, 197
Hittites, weaponry, 157
Hong Kong, relationship with China, 102
Hong Kong Exchanges and Clearing (HKEx), 101–2, 253n21
Hotel e Termas de Araxá, 38

"How Forward Integration along the Rare Earth Value Chain Threatens the Global Economy" (Boogaart), 34
Hudson Metals, 93
Hunter, Duncan, 28–29
Hydraulic mining, 158

IEA (International Energy Agency), 124–25, 136, 208, 228–29
IFixit, 216
Illegal mining and trading, 102–12
Incandescent bulbs, 150
Incentives, for rare element production, 226
India: energy demand in, 208; recycling in, 191; steel production in, 64
Indium: characteristics of, 3; pricing of, as by-product production, 80; processing of, 78; telemarketing sales of, 251n7; trading of, 97, 103, 205; uses of, 2, 13, 123, 187, 264n33
Indonesia: defense expenditures, 278n31; illegal minerals trade in, 105–8; social media use in, 126–27
Industrial accidents, 70, 81
Industrial products, resource demands for, 179
Industrial recycling, 185–86
Infrastructure, technological innovation in, 217–18
Inner Mongolia, export controls supporting, 202
Inner Mongolia University of Science and Technology, 196
Innovation distortion, 140, 154

Integrated circuits, 117–18
Intel, 8, 168, 214
IntelliMet, 70
Intercontinental ballistic missiles, 279n33
Intermetallics, 205
International Energy Agency (IEA), 124–25, 136, 208, 228–29
International Materials Agency, need for, 229
Internet cafés, 126, 127
IntierraRMG, 51
Investments, in rare metals mining, 49–54
InvestorIntel Technology Metals Summit, 50–51
Investor types, 60–61
iPhone, 1–3, 10
Iridium, 144
Iron, 13, 20–21, 26, 29, 57, 71, 78, 157–58, 163, 176, 178, 189, 197, 200, 235n6, 294n14
Iron Age, 12, 13, 157
Iron Dome (Israeli weapon system), 13

Jaffe, Robert, 148, 208–9, 210, 212–13
Jaffe, Sam, 151
Jakarta, Indonesia, construction in, 10–11
Japan: CBMM ownership in, 42; China and, x, 15, 22–25, 36, 165; government policies, effects of, 227–28; minor metals trading in, 89–90; Osaka, pollution in, 181; rare metal security strategy, 203–5, 212; recycling possibilities in, 187; U.S. embargo against, 30

Japan Institute of Metals, 219
"Jesus Phone," 1
Jet engines, 128
Jiangxi, China: ore processing in, 77, 82–85; pollution in, 173–75
Jiangxi Rare Earth Association, 198
Jobs, Steve, 1, 2, 3, 9
Johnson, Clarence "Kelly," 155, 158
Johnson Matthey, 186
Junior mining companies, 49–54, 59. See also Avalon Rare Metal

Kazakhstan, rare metals production, 113, 256n42
Kazatomprom, 113, 256n42
Kenya, Tiomin Resources and, 46–48
Kilby, Jack S., 117–18, 258n6
Kingsnorth, Dudley, 56
Kirby, Mary, 122–23
Kirchain, Randy, 110, 131, 132
Krupp, 160–61

Labeling, for rare metals resource efficiency, 224–25
Lache, Rod, 145
Lanthanum, 74, 75
Large Hadron Collider (CERN), 81
Latiff, Robert, 166, 169, 172
Lazer Tag (game), 120
LCD screens, 187, 264n33
Lean supply chains, 130–31
LEDs (light-emitting diodes), 122, 150–51
Lehrman, Danny, 63–64, 93–94, 95–96, 114
Lehrman, Noah, 91–93

Lengshuijiang, China, antimony
    production in, 40
Leveque, Alain, 70
Liberal Democratic Party
    (Japan), 23
Light-emitting diodes (LEDs),
    122, 150–51
Lighting and lightbulbs, 8, 35,
    150–52
Light rare earths, 49, 75, 227
Li Guirong, 175–76
Li Keqiang, 36
Lin, Maya, 8
Lithium, 2, 75–76, 116, 147, 168,
    189, 209, 219, 285n33
Lithium-ion batteries, 131, 146, 147,
    285n33
Lockheed Martin, 155–57, 158–59
London Metal Exchange (LME),
    101–2, 253n21
Long-term contracts, 96
Lovallo, John, 270n30
Low-carbon technologies, 179,
    294n14
Lubett, Luf, 234n2
Lumley, Graham, 58
Lynas Corporation, 183, 205

MacDonald, Ron, 50–52
Magnets: in cars, 141–43, 146;
    cerium in, 140–41; cobalt in,
    235n5; composition of, 235n6;
    decline in use of, 296n26;
    dysprosium in, 4, 21–22, 25,
    27–28, 205, 220, 229, 267n9;
    electrons in, 235n8; elements in,
    235n6; in elevators, 150;
    gadolinium in, 221; in genera-
    tors, 266n7; infrastructure use
    of, 217; military use of, 169;
    penny magnets, 261n16;
    permanent magnets, 3, 18–22,
    25–28, 141, 237n21, 268n14,
    296n26, 297n33; producers of,
    260n16; rare earth magnets,
    112–13, 122, 165–66, 186, 191, 211,
    230; rare earth permanent
    magnets, 137–38, 139, 145–46;
    reduced use of rare metals in,
    229–30; samarium-cobalt
    magnets, 20, 22, 190; strength
    of, 235n8; in toothbrushes, 116.
    See also Neodymium-
    dysprosium magnets
Malaysia: opposition to mining
    in, 49; rare earth production
    in, 49
Manhattan Project, 164
Manmade ores. See Recycling
Manufacturing, material loss
    during, 186
Marathon Oil, 87
Market data, need for, 226
Market for rare metals. See
    Trading networks
Market mining, 56–59
Massachusetts Institute of
    Technology (MIT), 136
Material Policy Commission, 30
Material recycling, 184–92
Material scientists, lack of, 85–88,
    172, 222
Materials consumption, increase
    in, 215
Materials Research Society, 135,
    208
Material supply, Graedel on,
    234n19
Mayer-Schönberger, Viktor, 119
McCallum, Bill, 140–41

McClean, Bill, 131
McKinsey consultancy company, 146
Medical implants, 163
El Mercurio (newspaper), on illegal mining, 112
Merrill, Charles, 29
Metallurgists, 85–88, 93, 172, 221–23
MetalMiner, 147
Metal-Pages, 94
Metals. *See* Rare metals; *names of specific metals*
Metcalfe, Robert, 220
Metzger, Wyatt, 148, 149
Microchips, 117–18
Middle class, projected increase in, 219
Military (general). *See* Wars
Military (U.S.): germanium use, 164; materials supply chain, 14, 167–68; rare metals use, 210
Miller, John, 141
Mineral deposits, formation of, 57
Minerals, Critical Minerals, and the U.S. Economy (National Academy of Sciences), 128
"Minerals and Metals Scarcity in Manufacturing: It Is the Ticking Time Bomb" (PwC), 131
Ming (pseud., exchange executive), 98–100, 102
Mining: ancient Roman techniques, 158; energy use in, 283n19; environmental impact of, 179–80, 181–82; government regulation of, 225; green economy and, 135; junior mining companies, 49–54, 59; mining companies, challenges

faced by, 46–54; opposition to, 40, 49; resource security and, 209; workforce age, 85
Ministry of Industry and Information Technology (China), 198, 282n8
Minor metals, 232n7. *See also* Rare metals
Minor Metals Trade Association (MMTA), 5, 6–7, 92, 95
MIT (Massachusetts Institute of Technology), 136
Mitsubishi Corporation, 113, 182–83
Mobile phones, 120–21, 179, 187, 260n14
Mobutu Sese Seko, 18, 20
Molybdenum, 29, 30, 49, 101, 148, 160–63, 211, 240n34, 250n20, 251n3, 276nn20–21
Molycorp, 74, 75, 77, 113, 184, 260–61n16
Monazite, 48, 105–6, 182–83
Monopolies, on production, 40–41
Moore, Kevin, 132, 142, 143
Moreira Salles family, 41
Morgan Stanley, 101
Motorola, 120–21
Mountain Pass mine, 77
"Mr. Rare Earth" (Karl Gschneidner), 28, 130
M2 Bradley Fighting Vehicle, 169
Multi-touch glass screen, 1
Musk, Elon, 223

Nakamura, Shigeo (Super Mario), 89–90, 113–14
National Academy of Sciences, 128
National Intelligence Council, 219

National Materials Advisory Board (U.S. National Research Council), 41

National Renewable Energy Laboratory, 217–18

National struggles, 18–37; China, control of rare earth element supply chain, 32–37; China-Japan conflict, 22–25; Congo, 20–22; Sagawa's permanent magnets and, 25–28; U.S. resource security and, 28–32; Zaire, 18–20. See also Geopolitics

Natural resources, growing demand for, 31–32

Nazis, metals use, 30

Neodymium, 4, 20–21, 26, 35, 70, 74, 89, 116, 135, 136, 140, 151, 190

Neodymium-dysprosium magnets, 4, 21, 26, 140–41

Nickel: in batteries, 116, 131; in magnets, 235n6; processing to produce cobalt, 78

Niobium, 3, 39, 41, 43–46, 52, 55, 56, 57, 64–66, 67, 68, 80–82, 89, 135, 148, 152–53, 207, 247n52

NioCorp (Quantum Rare Earth Developments), 56, 66

Nokia, 178

Nuclear industry, U.S. stature in, 212

Nuclear-powered submarines, 167–68

Obama, Barack, 209

O'Brock, David, 73–74, 80–81, 88, 103

Oil pipelines, 52

Oil shale, 134

Okabe, Toru, 221

Olds, Ransom E., 143

Olsen, Ken, 220

Optical fibers, 124, 262n23

Oral health. See Toothbrushes; Toothpaste

Origins of rare metals, difficulty of determining, 14–15

Osaka, Japan, pollution in, 181

Osram (lighting manufacturer), 151

Outsourcing, metals sources and, 110

"Oxcart" spy plane (A-12, "Titanium Goose"), 155–57, 158–59

Paint manufacturing, 19

Pake, George, 214

Palladium, 76, 144–45, 269n27

Paperless office, 214–15, 292n1

Patents, xi, 143, 211

Paul, Scott, 167

Penny magnets, 261n16

People's Bank of China, 101, 102

Periodic table, 4, 170–71; war over, 194–213. See also Geopolitics

Permanent magnets, 3, 18–22, 25–28, 141, 237n21, 268n14, 296n26, 297n33

Perskovites, 296n26

Personal consumption, increase in, 215

Peru, illegal mining in, 112

Philips Corporation, 110

Phillips, Jeff, 56, 61–62

Phones: iPhones, 1–3, 10; mobile phones, 120–21, 179, 187, 260n14; rare metals in, 27; recycling of, 224; smartphones, 121, 216, 218, 260n15

Pietrobono, Amber, 120
Planes, 128–31, 155–60, 168, 274n6,
    279n33
Planned obsolescence, 216
Platinum group metals, 144–45,
    178, 186, 249n13
PlayStation 2, 131
Political unrest, 48. *See also*
    Conflicts, funding of, from rare
    metals production
Politico, on military supply lines,
    167
Pollution, 153, 173–77, 181, 182–83.
    See also Carbon emissions
Post-consumer recycling, 186–87
Potvin, J. C., 46–47
Power packs, vanadium in, 52
Praseodymium, 74
Precious metals, 4, 190. *See also*
    Gold; Silver
Prices: benchmark prices, lack of,
    94–95; price bubbles, 113–14;
    rare earth crisis, 138–40, 227;
    volatility of, 90–91
PricewaterhouseCoopers (PwC),
    51, 131, 138
Processing industry, 276n20
Production (of rare metals):
    change in quantities of, 8–9;
    environmental impact of, 16,
    177–79; lead time for, 49–50;
    product losses during, 76–80,
    283n13
Production difficulties, 67–88;
    Chinese refining capacity,
    82–85; element isolation, 72–76;
    overview, 67–69; production
    efficiencies and costs, 76–80;
    refining, steps in, 69–72;
    Silmet's niobium processing,

80–82; workforce, lack of
    knowledgeable, 85–88
Product lifecycles, 178
Promethium, 72
Purdue University, 164–65
PwC (PricewaterhouseCoopers),
    51, 131, 138

Qinhua Wang, 203
Quantum Rare Earth Develop-
    ments (NioCorp), 56, 66

Radar, 164–65
Radiation-detection systems, 167
Radioactive materials, 3, 38, 40,
    55, 71, 176, 179, 182–83
Ralston, Oliver, 276n20
Rappaport, Michael, 94
Rare earth crisis, 138–40, 227
Rare earth elements (REEs), xi, 5,
    61, 202
Rare earth magnets, 112–13, 122,
    165–66, 191
Rare Earth Office, Ministry of
    Industry and Information
    Technology (China), 198
Rare earth permanent magnets,
    137–38, 145–46
Rare Metal Age, ix, 3, 120, 130,
    214–30; business models, need
    for change in, 223–25; conclu-
    sions on, 229–30; consumer
    habits, need for change in, 223;
    decision-making processes,
    227–28; in developing coun-
    tries, 218–19; global forum,
    need for, 228–29; government
    role in, 225–27, 228; from
    National Renewable Energy
    Laboratory, 217–18; overview,

214–17; rare metals supply and, 219–20; research, need for, 220–23; smartphones, 218
Rare metal exchanges, 96–98
Rare metal risk committees, 131
Rare metals: corporate issues, 38–66; environmental needs, 134–54; geopolitics, 194–213; list of, 6–7; national struggles and, 18–37; overview, 1–17; production difficulties of, 67–88; Rare Metal Age, prospering in, 214–30; sustainable use and, 173–93; tech needs and, 115–33; as term, xiii, 232n7; trading networks for, 89–114; ubiquity, 3; in war, 155–72. *See also names of individual elements*
Raytheon, 171
Reagan, Ronald, 30, 239n30
Reck, Barbara, 188, 189, 190, 191, 192
Recycling, 177, 184–92, 224, 285n34
Reeck, David, 141
REEs (rare earth elements), xi, 5, 61
Refining, steps in, 69–72
Reisman, Lisa, 110, 111
Renewable energy, 135, 136–37
Research, need for, 220–23
Reshoring, 212
Resource constraints, 205–8
Resource needs, x, 12
Resource security, 28–32, 203–5, 208–12, 228
Rhenium, 113, 114, 128–30, 132, 256n42
Rhodia (Solvay, metal processing firm), 70

Richard Hammen, 70
Richardson, Ed, 165–66
Rive, Lyndon, 148
Romans (ancient), 158
Roosevelt, Franklin Delano, 30, 238n28
Royal Philips Sonicare, 115
Rule of law, 200–201

Sagawa, Masato, 20–22, 25–28, 205, 229
Samarium, 20, 121
Samarium-cobalt magnets, 20, 22, 189–90
Santini, 235n2
Saudi Arabia, control of oil prices, 40
Savagian, Pete, 142, 143
ScanWind, 139
Schlumberger, 86
Screens: flat-screen technology, 122–24; LCD screens, 264n33; touch screens, 1, 261n19
Seafood, 208–9
"Sea of Death" (Dokai Bay, Japan), pollution in, 181
Search engine use, in China, 199
Secondary sources, rare metals recovery from, 79–80
Second China Rare Earth Summit, 194
Secrecy, xi–xii, 67, 68, 70, 87, 147, 160, 161, 196, 197, 268n17
Securing Materials for Emerging Technologies (American Physical Society & Material Research Society), 208, 211, 212–13
Selenium, 79, 149, 150, 190, 246n40
Semiconductors, 165

Senkaku Islands (Diaoyu Islands), territorial dispute over, 22–24
Senter, William, 156
Sesa-Opas (pseud., Indonesian social media user), 126, 127
Shanghai International Securities, 252–53n17
Shape memory, 221
Sheffi, Yossi, 112–13, 172
Sichuan Province, China, opposition to mining in, 49
Siemens, 139
Sillamäe, Estonia, refining in, 67–69
Silmet (Factory Number 7), 67–69, 72–76, 80–82
Silver, xiii, 4, 187, 190
Silver, Michael, 35, 139–40, 218
Simbol Materials, 211
Sinclair, Clive, 118, 119
Skyscrapers, molybdenum in, 162
Smartphones, 121, 216, 218, 260n15
Smith, John, 131–32
"Smoke Capital" (Osaka, Japan), pollution in, 181
Smuggling, 98, 203. See also Illegal mining and trading
Social media, mining company investments via, 51
Solar panels, 148–50, 152
Solid-state drives, 122
Solvent extraction, 70
Solyndra (solar panel manufacturer), 209
Sonneborn, Jon, 145–46
South Korea: CBMM ownership in, 42; steel demand in, 11
South Portland, Maine, landfill mining in, 224

Soviet Union, cobalt purchases by, 18–19, 234–35n2
Spurr, Josiah, 291n33
Sputtering (manufacturing process), 185
Spy planes, 155–57, 158–59
Stainless steel, 74, 155–56
Stalin, Joseph, 67, 68
Standard of living, 10–11
Start-stop batteries, 147–48
State Council (China), 198
Statistics on rare metals industry, lack of reliable, xii. See also Secrecy
Steel: airplanes and, 128; automobiles and, 131, 276n20; consumption of, 11; demand for, 58, 152–53; military use of, 160–62; pollution from processing of, 181; processing of, 190; quality in China and India, 64; recycling, 185; strengtheners for, 29, 153; strength/toughness of, 44–46, 52, 156; titanium compared to, 221
Stockpiling, 226–27
Strangas, Elias, 296n26
Strategic metals. See Rare metals
Stratton, Patrick, 108
Strontium, 121
Submarines, nuclear-powered, 167–68
Substitute performance, 169, 170–71
Sunny (Chinese consumer), 125
Super Mario (Shigeo Nakamura), 89–90, 113–14
Supply and demand, imbalances in, 15–16, 132–33

Supply lines, x–xi, 29–30, 154–55,
    166–69, 219–20
Sustainable use, 173–93; conclu-
    sions on, 192–93; Dalahai,
    China, pollution in, 175–77;
    environmental regulations,
    compliance with, 184; Jianxi,
    China, pollution in, 173–75;
    material recycling, 184–92;
    mining, environmental impact
    of, 179–80, 181–82; Osaka,
    Japan, pollution in, 181;
    production, environmental
    impact of, 177–79; radioactive
    materials, pollution from,
    182–83. See also Environmental
    needs
Sykes, John, 58
Sylvania, 151

Tablets (electronic), 120, 218
Tailings (mining waste), 175, 180
Talent drains, problem of, 85–88
Tantalum, 2, 48, 57, 67–68, 80,
    89, 108–9, 111, 116, 121, 124, 131,
    168, 185
Task Force on American Innova-
    tion, 163
Taylor, Patrick, 55, 222
Tech needs, 115–33; aviation
    sector, 128–31; in China, 125;
    conclusions on, 132–33; fiber
    optics, 124; flash drives, 121;
    flat-screen technology, 122–24;
    in Indonesia, 125–28; integrated
    circuits, 117–20; laptops, 121;
    mobile phones, 120–21, 179, 187,
    260n14; overview, 115–17;
    resource scarcity, potential for,
    131–32

Technologies: expectations of,
    9–10; predictions on, 220. See
    also individual technologies
    (e.g., batteries, wind turbines)
Technology metals. See Rare
    metals
Teck Resources, 184
Teddy Ruxpin (talking teddy
    bear), 119
Telemarketing, indium sales via,
    251n7
Tellurium, xiii, 78–79, 80, 148–50,
    167, 190, 207, 209, 246n40
Tenent, Robert, 217
Teng Biao, 200
Teran, Alex, 147
Terbium, 2, 4, 151, 167, 174, 206, 229
Territorial disputes, China-Japan,
    22–24
Tesla, 145–47
Texas Instruments, 117, 118
Thatcher, Margaret, 30
Thermal-imaging systems, 163–65
Thin film technologies, 148–49
Thorium, 3, 57, 176
Thor Lake mine (Avalon Rare
    Metal), 54–56
3D printing, 221
Tin, 48, 105–7, 108
Tiomin Resources, 46–48, 54
Titanium: in aerospace industry,
    162, 263n30; in airplanes, 96,
    128, 156–60, 168, 274n6;
    commoditization of, 221;
    market for, 44; in mobile
    phones, 121; nonmilitary use of,
    121, 124, 146, 162–63, 221; shape
    memory, 221; sources of, 46, 93,
    113; U.S. research on, 206; in
    weaponry, 163, 168, 169

"Titanium Goose" (A-12 spy
plane, "Oxcart"), 155–57, 158–59
Titanium Metal Corporation, 156
Toothbrushes, 115–17, 258n3
Toothpaste, ancient use of,
257–58n1
Toronto Stock Exchange, 50
Toshiba, 112–13, 256n42
Touch screens, 261n19
Toyota Tsusho, 108
Toys, electronic, 119–20
Trade secrets, xi
Trading networks, 89–114; China,
rare metal exchanges in, 96–98;
China, regulatory environment
in, 98–101; conflict funding,
108–12; evading export
controls, 104; export quotas/
ban set by China, x, 24, 240n34;
Indonesia, illegal trade in,
105–8; Lehrman family, 91–96;
limited suppliers, problem of,
112–13; London Metal Ex-
change, 101–2; overview, 89–91;
precariousness of, x; price
bubbles, 113–14; secrecy in, 16;
smuggling, 102–5
Truman, Harry S., 30
Tungsten: Allied actions on, in
WWII, 239n28; China,
production in, 32, 205, 240n33,
289n16; conflict tungsten, 108,
109; Congo production, 108;
export quota, 240n34; in glass,
217; importance, xi, 11; in
lighting, 151; patents, 211;
production locations, 48;
shortage fears, 207, 219; sources
of, 32, 48, 93, 108, 205, 240n24,
240n33, 289n16; wartime use of,
29, 30, 239n28; in weaponry, 29,
161–62, 167
Tunna, Nigel, 96
Twitter, 126

Uganda, cassiterites from, 111
Umicore, 191
UN Intergovernmental Panel on
Climate Change, 135
United States: aluminum can
recycling, 285n34; Bureau of
Mines closure, 222; China,
trade case against, 36; on
China's materials exports, 203;
cobalt supplies, 19; commodity
stockpiles, 291n36; conflict
materials, actions on, 110–11;
Japan, embargo against, 30; rare
metal security strategy, 206,
208–12; reshoring, 212; tungsten,
wartime actions on, 162, 239n28.
See also Military (U.S.)
Uranium, ix, 67–68, 164
Urban mining. See Recycling
U.S. Chamber of Commerce,
288n11
U.S. Commercial Company,
238n28
U.S. Defense Department: rare
metal stockpiles, 31. See also
Military (U.S.)
U.S. Department of Energy,
136–37, 138, 146, 207
U.S. Geological Survey, 91
U.S. National Renewable Energy
Laboratory, 79, 80
U.S. National Research Council,
National Materials Advisory
Board, 41
U.S. Navy, 235n5

Vale Fertilizantes, 39
Value chain for rare earth
    elements, 34–35
Vanadium, 29, 30, 50, 52, 59
Vestas, 139
Vietnam: defense expenditures,
    278n31; illegally mined rare
    metals in, 103
Vries, Eize de, 139
VSMPO-AVISMA (titanium
    manufacturer), 113

Wadia, Cyrus, 146–47
Walsh, David, 59
Wang (pseud., metals trader),
    100–101
Wang Shangkun, 125
Wars, 155–72; germanium use in,
    163–65; high-performance
    materials, need for, 169, 171–72;
    historical overview, 157–58;
    impact on nonmilitary uses,
    162–63; materials research,
    166–67; "Oxcart" spy plane,
    155–57, 158–59; resource supply
    lines in, 29–30, 167–69;
    tungsten use during, 29, 30,
    239n28; WWI, 160–62
Wars over periodic table, 194–213.
    See also Geopolitics
Water purity, 284n25
Watson, Emma, 126
Weapons, 13–14, 29–30, 157–58,
    160–62, 167–69
Wearable electronics, 126

Wedding bands, resource
    demands for, 178
Windows, energy efficiency of, 217
Wind turbines: Alstom on, 266n7;
    in China, 35–36, 154; drive
    trains for, 137–39, 266n8,
    268n14; in Estonia, 134–35
Winergy, 268n14
Wireless networks, 124
Workforce, lack of knowledgeable,
    85–88
World Trade Organization
    (WTO), 36, 200–203
World War I, rare metals use
    during, 160–62, 207
World War II, 204; tungsten,
    Allied actions on, 239n28

Yang, Jiang, 97–98, 101
Yang Guohong, 97
Ytterbite (gadolinite), 72
Yttrium, 11, 72, 128, 167
Yudi (pseud., metals trader),
    105–7
Yue, Steve, 85–86

Zaire, cobalt production, 18–20
Zandkopsdrift mine, 77
Zanuck, Darryl, 220
Zhang Hongjiang, 196
Zhang Yang'e, 173, 175, 178
Zhengzhou Commodity Ex-
    change, 100
Zhu Rongji, 200
Zinc, 4, 78–79, 177, 181